HERBIVORES

Edited by **Vonnie D. C. Shields**

Herbivores
http://dx.doi.org/10.5772/62658
Edited by Vonnie D. C. Shields

Contributors

Jane Williamson, Jacqueline Bell, Carolina Pietczak, Lucas Vieira, Alina Morquecho-Contreras, Hermilo Sánchez-Sánchez, Khalil Safaei, WenZhu Yang, Luis Ferreira, Rafa Celaya, Ana Santos, Miguel Rodrigues, Koldo Osoro, Kenneth J. W. Moore, Sotirios Archontoulis, Andrew W. Lenssen, Ute Wittstock, Maike van Ohlen, Anna-Maria Herfurth

Published by InTech
Janeza Trdine 9, 51000 Rijeka, Croatia

Publishing Process Manager Dajana Pemac
Technical Editor SPi Global
Cover InTech Design team

First published March, 2017
Printed in Croatia
Legal deposit, Croatia: National and University Library in Zagreb

Additional hard copies can be obtained from orders@intechopen.com

Herbivores, Edited by Vonnie D. C. Shields
p. cm.
Print ISBN 978-953-51-2987-5
Online ISBN 978-953-51-2988-2

For EU product safety concerns:
IN TECH d.o.o., Prolaz Marije Krucifikse Kozulić 3, 51000 Rijeka, Croatia,
info@intechopen.com or visit our website at intechopen.com.

Meet the editor

Vonnie Shields, PhD, is currently a full professor in the Biological Sciences Department and an associate dean in the Fisher College of Science and Mathematics at Towson University, Towson, MD, USA. Dr. Shields' laboratory engages in multidisciplinary research directed toward exploring the importance of gustatory, olfactory, and visual cues in the selection of food sources by carrying out behavioral and electrophysiological studies on larval and adult insects. In addition, her lab examines the structural organization of insect sense organs using transmission electron and scanning electron microscopy. The overall goal of this research is to acquire a better understanding of the sensory mechanisms by which insects find host plants and detect plant-associated volatiles. The aim is to discover possible novel biocontrol agents against insect pests. Dr. Shields studied biology at the University of Regina, Regina, Saskatchewan, CA. Her interest in insect chemosensory research began after her undergraduate studies, when she started her PhD studies at the same institution. For her PhD degree, she carried out research at the University of Regina and the University of Alberta, Edmonton, Alberta, CA. After graduating, she accepted a research associate position to conduct postdoctoral studies at the Arizona Research Laboratories Division of Neurobiology, University of Arizona, Tucson, Arizona, USA, before she accepted a faculty position at the Towson University.

Contents

Preface

Most simply defined, herbivory is generally described as the consumption or predation of plant matter by animals or organisms physiologically adapted to consuming plants, algae, and photosynthesizing bacteria. This book contains chapters from a variety of topics that fall into the following broad sections: (I) "Plant Defense Mechanisms and Herbivore Adaptations," (II) "Herbivory and Food Processing of Grazing Animals," and (III) "Herbivory Effects on Plant Communities." This book presents comprehensive reviews focusing on defense mechanisms of plants against herbivores or herbivore adaptations to overcome plant defenses, diet selection by evaluating diet composition of ruminants, modeling ruminant herbivory, the effect of grain processing on digestion and overall health of cattle, and marine herbivory. The chapters in this book are written by experts in their respective fields and will be a valuable resource for general biologists as well as botanists, ecologists, and zoologists, and students training in these areas.

In Chapter 1, authors review how plant defense chemicals, activated upon herbivory, reduce the ability of herbivores to obtain plant nutrient tissues. The authors discuss how such defenses induce a complex network of both direct and indirect defenses. The former directly influence the herbivore, while the latter, such as volatile emissions produced as a result of herbivore attack, function as attractants for other potential insects to, in turn, predate on the herbivore. All of these interactions, with respect to population dynamics and evolution in both plants and invertebrates, are discussed.

In Chapter 2, authors discuss how many plants have developed chemical defenses to ward off herbivores. An example of this is cyanide, an ubiquitous plant-produced compound, stored as cyanogenic glucosides and used by the plant as potent antiherbivore defenses. Once the plant tissue is disrupted, due to cyanogenic glucoside hydrolysis ("cyanide bomb"), cyanide is mobilized. As this compound is an inhibitor of cellular respiration, it is an acute toxin for all aerobic organisms. In order to use cyanide-containing plants as food sources, herbivores have evolved mechanisms to overcome these defenses to overcome cyanide defenses of their host plants. This chapter presents the current understanding of cyanide detoxification pathways and the enzymes involved, as well as cyanide avoidance mechanisms in herbivores, specifically arthropods. Cyanogenesis in herbivores is discussed, in addition, from both ecological and evolutionary perspectives.

In Chapter 3, authors investigate conventional as well as newer methodologies for studying diet selection of domestic and herbivores under diverse conditions. More specifically, they focus on the utilization of epicuticular compounds, namely, alkanes (i.e., long-chain fatty acids as well as alcohols), as fecal markers. Such studies will help with the development of appropriate grazing strategies to create more efficient and sustainable management and production systems for vegetation communities and to increase our knowledge of grazing behavior.

In Chapter 4, authors review how relatively simple models can be used to describe important nutritional processes related to ruminant herbivory. More specifically, the authors utilize models that describe the relationship between rumen escape protein and protein concentration, kinetics of fiber digestion, true digestion, and potential intake of herbage.

In Chapter 5, authors examine the improvement and optimization of grain digestion in ruminants. More specifically, the effect of the grain processing method, the degree of processing on rate and extent of grain digestion, the effects on lactation performance, and cattle health are considered.

In Chapter 6, authors explore the importance of direct and indirect interactions involving marine herbivores and algae yielding positive and negative results. An understanding of these interactions is essential in predicting accurately the impact of potential perturbations for successful management of ecosystems.

In Chapter 7, authors address herbivore consumption by lizards, a topic often overlooked. They consider how consumption of strictly plant matter by lizards occurs in only a relatively small percentage, which is typically lower and less frequent than that of other animal groups (e.g., mammals and birds), as omnivorous species tend to eat more fruits, flowers, and nectar. The latter food products are easier to digest and provide more nutrients than leaves. Included in the chapter is a discussion of seed dispersion and flower pollination by lizards as well.

I wish to thank InTech Open Access Publisher for initiating this book project and inviting me to serve as editor. I would like to recognize the Publishing Process Manager, Dajana Pemac, assigned to the task of publishing this book, for guiding me through the process. I would like to acknowledge all the authors for their hard work in submitting and editing their contributions. Lastly, I wish to express a special thanks to my husband, Dr. Thomas Heinbockel, Professor and Director of Graduate Studies, Department of Anatomy, Howard University College of Medicine and to our son, Torben Heinbockel, for their patience and understanding in the last year when I was working on this book project.

Dr. Vonnie D.C. Shields
Associate Dean, Jess & Mildred Fisher College of Science and Mathematics
Professor, Biological Sciences Department
Towson University, Towson,
Maryland, USA

Section 1

Plant Defense Mechanisms and Herbivore Adaptations

Chapter 1

Chemical Plant Defense Against Herbivores

Hermilo Sánchez-Sánchez and

Alina Morquecho-Contreras

Additional information is available at the end of the chapter

http://dx.doi.org/10.5772/67346

Abstract

Herbivores can damage plant productivity and fitness because plants have improved defense mechanisms such as physical barriers, association with other organisms such as ants, and chemical defense. In that, separate plant species produce different chemical molecules. Chemical compounds involved in plant defense can act in several facts: decreased palatability, like a poison, such as a stunner, and increased gene defense expression, among others. In this chapter, we approach several examples of chemical molecules produced by plants to defend themselves, including biochemical metabolic pathways, as well as ecological and evolutive implications.

Keywords: plant chemical defense, alkaloids, resins, herbivores

1. Introduction

Interactions between plants and insect herbivores are important determinants of plant productivity in managed and natural vegetation. In response to attack, plants have evolved a range of defenses to reduce the threat of injury and loss of productivity [1]. Plants are exposed to threats of resource loss by herbivory in natural conditions experiencing damage; to mitigate losses many plant species develop defensive traits against herbivores, such as primary and secondary metabolites [2–4]. Among herbivores are many arthropods, mollusks, vertebrates, and nematodes, and these groups consume between 5 and 20% of plant biomass annually [5].

The cost on investing in defense can be quantified in reduced growth, lower photosynthetic production, and reduced plant fitness [6, 7]. Plant defenses reduce the ability of herbivores to obtain nutrients from plant tissue. Plants with diminished defense capability may suffer greater herbivore damage and exhibit lower overall fitness under conditions of herbivore stress than well-defended plants [7].

Plants respond to herbivory through various morphological, biochemical, and molecular mechanisms [8] and exhibit multifactorial traits against herbivory that are constitutively expressed or induced upon attack [9]. The plant defense activated upon herbivory is a complex network of different pathways composed of direct and indirect defenses. Direct defense compounds such as glucosinolates or protease inhibitors directly influence the insect performance and feeding behavior, while indirect defenses like emission of volatile organic compounds after herbivore attack function as attractant for parasitic wasp which in turn predate on the attacker. While plants develop new defense compounds or mechanisms to enhance the resistance against herbivores, their attackers find new ways to bypass or detoxify these [8, 10].

Insect herbivory induces several internal signals from wounded tissues, including calcium ion fluxes, phosphorylation cascades, and systemic and jasmonate signaling. These are perceived in undamaged tissues, which thereafter reinforce their defense producing low molecular weight defense compounds [11]. Some compounds produced by plants constitutively or induced by herbivore damage are toxic or impair gut function in arthropod; examples include alkaloids, benzoxazinoids, glucosinolates, and terpenoids [1].

Added to this, there are some other defense mechanisms, such as mechanical defenses, indirect defenses, interactions with other organisms, etc. In this review, we focus in different traits defensive in plants and its effect on population dynamics and evolution in both plants and invertebrates. Finally, we integrate all traits in a specific example in *Pinus* genera.

2. Induced defenses

Plants respond to herbivore attack through a dynamic defense system that includes structural barriers, toxic chemicals, and attraction of natural enemies of target pests. Both defense mechanisms may be present constitutively or induced after damage by the herbivores. Most of chemicals are produced in response to herbivore attack. Induced defenses make the plants phenotypically plastic, and high variability in defensive chemical exhibits a better defense [8].

The induced defenses occur when past or current herbivory is a reliable cue of future attack and defenses are costly; while in environments where herbivory is constantly high, constitutive defenses should be favored [4].

Herbivorous insects produce oral secretions which contain compounds that elicit plant responses [12] and plant elicitor peptides prevalence across wide-ranging plant families [13]. In response, plant produces diverse chemical active compounds such as benzyl cyanide, fatty acid-amino acid conjugates, and proteins such as β-glucosidase [14]. Plants can recognize herbivore elicitor and initiate a cascade of responses, including changes in plasma membrane potential and activation of networks of kinases and phytohormones [15]. Three major plant hormones, jasmonic acid (JA), salicylic acid (SA), and ethylene (ET), function in a complex regulatory network essential in herbivore-induced defense responses [16].

http://dx.doi.org/10.5772/67346

3. Chemical compounds in plant defense

Plants produce defensive metabolites, which do not affect the normal vegetative growth and development, but reduce the palatability of tissues in which are produced. Can be constitutive stored as inactive forms or induced in response to insect or microbe attack [8]. The defensive metabolites are bioactive specialized compounds used to protect plant against herbivores, and these compounds can use as target systems unique to herbivores, such as the nervous, digestive, and endocrine organs, may act as repellents for generalist herbivores, while specialists are forced to invest resources in detoxification mechanisms [11, 17].

Plant defense include changes in transmembrane potential immediately upon herbivory damage and are tightly followed by changes in the intracellular Ca^{2+} concentration and generation of H_2O_2. Kinases phytohormone jasmonic acid (JA), ethylene (ET), salicylic acid (SA), and nitric oxide (NO) are detectable within minutes. After roughly 1 h, gene activation is followed by metabolic changes [11, 13].

Antinutritive proteinase inhibitors (PINs) are locally and systemically induced upon insect attack, but many other proteins contribute to antiherbivory responses. Enzymes such as polyphenol oxidase a threonine deaminase limit protein availability in the midgut, whereas others destabilize insect peritrophic membranes [13, 18]. Plants also draw upon a complex arsenal of small-molecule chemical defenses including terpenoids, alkaloids, phenylpropanoids, glucosinolates, lipids, and nonprotein amino acids [19].

Volatiles which can alert neighbor plants or tissues to potential attacks are promoted by herbivory and are a complex blend. Volatiles induces indirect defenses inhibits oviposition and attracts natural enemies such as parasitoids and predators [13].

3.1. Alkaloids

Efficient feeding deterrents against herbivore group of compounds are the alkaloids, particularly such derived from quinolizidine, like cytisine and sparteine. These molecules are alkaline and contain nitrogen in a heterocyclic ring [20]. Alkaloids are biosynthesized in roots from amino acids [21] and probably are involved in defense against insect herbivory. Twenty percent of vascular plants synthesized alkaloids, particularly in plant families Leguminosae, Liliaceae, Solanaceae, and Amaryllidaceae [11, 19].

3.2. Phenolics

Phenolics are produced by plants as compounds able to repel herbivores, inhibit enzymes, attract pollinators and fruit dispersers, absorb UV radiation, and decrease competition between plant neighbors [11, 22]. There are approximately 10,000 plant phenolics derived from shikimic y/o malonic acids [23]. Phenolics can bind covalently to herbivore's digestive enzymes and inactivate them [24] or halt the growth and development of larvae [25]. Phenolics can be regulated for external conditions like light and nutrients; when a plant is stressed, it produces less phenolics than nonstressed plants [22].

3.3. Terpenoids

The most diverse class of bioactive natural products in plants is terpenoids, with approximately 40,000 structures. Terpenoids are synthesized from acetyl-CoA and play a role in plant defense, can act like active compounds in resin or as volatiles, repellents, and toxins, or can modify development in herbivores [26]. Another characteristic in monoterpenes and sesquiterpenes is its ability to form essential oils, like limonene in citrus plants; these essential oils have repellent and toxic effects on insects [27]. Many terpenoids can have synergistic effects upon release [28].

3.4. Nonprotein amino acids

Amino acid g-aminobutyric acid (GABA), a four carbon nonproteinogenic widespread in animals, plants, and microorganisms, can be implicated in defense responses. Wounding plant tissue and cell disruption caused by feeding insects is sufficient to induce rapid jasmonate-independent GABA synthesis and accumulation. When ingested the elevated GABA levels become toxic for the insects. GABA is synthesized by decarboxylation of L-glutamate bay glutamate decarboxylases (GAD) in shoots and roots and is a component in a plant's first line of general, rapid defense against invertebrate pests [29].

One metabolite induced in plants is tyrosine, which can be redirected into other primary and secondary metabolites, and its accumulation in excess in young leaves may not be adaptive as they would persist once the leaf is full in size and protected by toughness [30]. In contrast to tyrosine, physiological constraints on catabolism may be selected against induction of phenolics and saponins [4]. When plants exceed the capacity to store constitutive secondary metabolites could avoid autotoxicity [31].

3.5. Sulfur

Sulfur is a crucial element for plants, determining plant development, maintenance, and resistance to environmental stress. Sulfur is taken up by plants as inorganic sulfate and incorporated in different sulfated metabolites including glucosinolates, selected flavonoids, phytosulfokines, and hormones by distinct pathways. Some sulfated metabolites function in plant defense against pathogens and herbivores such as defensin and thionin peptide, antimicrobial defenses with widespread distribution, whereas antifeedant glucosinolates are limited to the Brassicales order. *Bacillus subtilis* activates plant growth by producing IAA y/o gibberellins and emits volatile metabolites (VOCs), which can activate transcripts related to cell wall modifications, primary and secondary metabolism, stress responses, hormone regulation, iron homeostasis, and sulfur-rich aliphatic and indolic glucosinolates. Plants exposed to *Bacillus subtilis* with elevated glucosinolates exhibit greater protection against generalist herbivores. Then, plant-growth-promoting rhizobacteria can enhance plant sulfur assimilation and integrate in plant defense [32].

3.6. Lipids

Fatty acids (FAs) are essential macromolecules present in all living organisms, are the major source of reserve energy, are essential components of cellular membranes, and are implicates as signaling molecules, modulating normal and disease-related physiologies in microbes, insects, animals, and plants. In plants, fatty acids regulate salt, drought, heavy metal tolerance, and herbivore feeding, especially by JA is a FA derivate molecule [33]. In *Nicotiana attenuata* fatty acid-amino acid conjugates (FACs) in the herbivore *Manduca sexta* oral secretions are the major elicitors that induce herbivory-specific signaling [34]. FAs increased plant defense against pathogens and insects by stimulation of key short- and long-term regulatory process [35, 136].

Simulated herbivory dramatically increased salicylic acid-induced protein kinase (SIPK) activity and jasmonic acid (JA) levels in damaged leaves and undamaged systemic leaves, whereas wounding alone had no detectable systemic effects. The activation of SIPK and elevation of JA in specific systemic leaves increase in the activity of an important antiherbivore defense, trypsin proteinase inhibitor (TPI). Then, *N. attenuata* can identify FACs produced by herbivory in damaged leaves and activate MAPK and JA signaling for activated defenses [34].

Another lipids produced by plants are alkamides. Natural alkamides are often insecticidal [35, 36]. Chrysanthemum cultivars show a wide variation in degree of host-plant resistance to the western flower thrips *Frankliniella occidentalis*. Extracts of chrysanthemum leaves revealed the presence of an unsaturated isobutylamide, N-isobutyl-(E,E,E,Z)-2,4,10,12-tetradecatetraen-8-ynamide. Alkamides account for natural host resistance to thrips. The participation of alkamides in host resistance to insects can be due to their role as elicitors of plant defense responses. For instance, it has been reported that linolenoyl-L-glutamine, an amide produced in oral secretions of caterpillars, is able to induce the production of volatile chemicals from plants that attract predators and parasites of the caterpillar while it feeds [36, 37].

3.7. Jasmonic acid and ethylene

Jasmonic acid (JA) is an important regulator of defense responses against chewing insects, necrotrophic pathogens, and cell-content feeders such as spider mites and thrips [16]. Herbivores stimulate JA production by octadecanoid pathway. In *Arabidopsis*, JA is conjugated with isoleucine [135] through the enzyme jasmonoyl isoleucine conjugate synthase1 (JAR1) that conjugates binding to the F-box protein coronatine insensitive1 (COI1) and degrades jasmonate ZIM domain (JAZ) repressor proteins [38, 39]. Then, JA-responsive genes, including JAZ, which involves a negative feedback loop are activated [16]. There are two possible pathways: MYC2 regulates positively vegetative storage protein 2 (VSP2) and lipoxygenase 2 (LOX2), which are JA-responsive inducible by wound. The another pathway implicates the ethylene response factor (ERF) (JA and ET are synergic) and induces ERF1 and ORA59; both are JA/ET-responsive transcription factors which regulate responsive genes like plant defensin 1.2 (PDR1.2) [40]. MYC2 regulates defense against herbivores, and ERF is involved in induced defense especially against necrotrophic pathogens [16].

3.8. Salicylic acid

Salicylic acid (SA) is an essential signaling molecule that mediates pathogen-triggered signals perceived by different immune receptors to induce downstream defense responses. SA is a small phenolic phytohormone, which plays a major role in mediating defense; its accumulation is essential for induction of defense responses [40, 137].

Induced plant responses are regulated by SA when herbivores bite phloem [16]. Plant responses synthesizing SA from chorismate by isochorismate and phenylalanine ammonium lyase pathways [41]. Increases in SA concentrations lead to nuclear translocation of pathogenesis-related genes 1 (NPR1), which results in the expression of defense proteins, the pathogenesis-related (PR) proteins [42].

When a plant faces multiple herbivore attack, induced defense is regulated through interconnection of the JA, SA, and ET signal transduction pathways. Cross talk between JA and SA signaling is mutually antagonistic, resulting in the prioritization of SA-dependent defense responses over JA-dependent responses or vice versa [42].

4. Mechanical defenses

The first layer of defense in plant is mechanical, and the major components contributing to mechanical defenses are trichomes. These structures negatively influence on herbivore feeding behavior and insect mobility [43]. Another trait in plant defense is the palatability, and one form to modify this character is to produce dense trichomes; for example, in *Phaedon* species, the host preference of adult beetles was less for *Brassica* cultivars that produced dense trichomes, while adult beetles were inclined to attack glabrous leaves [3]. That is particularly important on young leaves of hairy plants, which produce denser trichomes than those of mature leaves. Therefore, trichomes might play an important role in the defense of younger leaves and contribute to future development of leaves [3, 44]. Trichomes tend to be more effective against insects that are small relative to trichome size; additionally, trichomes tend to deter sap-feeding or leaf-chewing insects to a greater extent than those feeding within plant tissues [45]. Spinescence, including spines, thorns, and prickles, also defends the plants against many insects [8].

Epicuticular waxes form a slippery film or crystals that prevent from attaching to the plant surface, oviposition, or feeding [1]. The biosynthesis and composition of waxes vary during plant development, and the physical-chemical properties of the cuticle respond on changes in season and temperature [46].

Another mechanical defense is to deposit granular minerals in tissues that deter insect attack and feeding. For example, Si accumulation, especially in Poaceae family, which is abrasive, damages herbivore feeding structures and reduces digestibility. Si accumulation can be induced by herbivory. Si in leaf surface can be abrasive in grasses with silicified spines, while others deposited Si in short cells. Si allocation to spines impacts palatability, while allocation to short cells may impact digestibility [1, 45].

The cell walls of leaves are also reinforced during the feeding through the use of different macromolecules, such as lignin, cellulose, suberin, and callose, together with small organic molecules, such as phenolics and Si [47].

http://dx.doi.org/10.5772/67346

Good few plants contain laticifers and resin ducts that canals produce and store latex and resins under internal pressure; when the channels are broken, they are secreted and might entrap or intoxicate the herbivore [11, 48]. However, several specialist herbivores can block the flow of latex cutting the leaf veins, for example, the milkweed beetles *Labidomera clivicolis, Tetraopes melanurus*, and *T. tetrophthalmus* for feeding *Asclepias* cut veins and wait stop flow [49].

Oleoresins produced by conifers are a blend of terpenoids and phenolics accumulated in intercellular channels. When bark beetles bite that channels resin flow and get out the insect until outside, when oleoresins solidifying [11, 50].

5. Indirect defenses

Indirect defense can be used when plants attract, nourish, or house other organisms to reduce enemy pressure [51]. For example, ant association in *Mallotus japonicus* (Euphorbiaceae) the damage leaf areas of ant excluded plants were much larger than those of control plants in middle-age leaves [44]. This is done by producing volatiles, extrafloral nectar, food bodies, and nesting or refuge sites [11].

Extrafloral nectar is secreted on leaves and shoots to attract predators and parasitoids and consists mainly of sugars, amino acids, lipids, proteins, antioxidants, and mineral nutrients; its production increases by herbivory and decreases in the absence of herbivory [52]. Extrafloral nectar has been associated to protective ants, which have the ability to defend their food sources. Increases in extrafloral nectar production augment the numbers of protective ants. In *Catalpa bignonioides* and Fabaceae family, extrafloral nectar attracts mites, ladybird beetles, wasp, lacewing larvae, and spiders [53].

6. Another influent factors in plant defense

The composition and dynamics of the insect community that interacts with plants are influenced by plant traits such as chemistry, physiology, and morphology, which have a genetic basis. Plant traits may affect the sizes of herbivores and therefore the sizes of parasitoids that develop in the herbivores and even the sizes of hyperparasitoids.

Induction of defense timing was examined by Bixenmann and collaborators [4] in *Inga* genus using lepidopteran larvae on young leaves. While young leaves are expanding, they are tender and high in protein, the two traits that make them a target for herbivores, receiving 70% of the leaf's lifetime herbivore damage despite being vulnerable for only few weeks. Once leaves reach their full size, they rapidly toughen, and rates of herbivore drop to almost zero. The amount of damage, the timing, and the identity of damage agent impact directly induced responses. When increasing leaf area removed in *Phaseolus lunatus*, extrafloral nectar production, and ant recruitment decreases significantly, then extrafloral nectar production is inversely correlated with leaf area and therefore with the amount of intact photosynthetic surface [54].

Herbivory risk depends not only on the traits of an individual plant but also on those of neighboring plants [3]. In that sense, the "associational effects" may mediate the local frequency of the density dependence of herbivory [55].

Volatile organic compounds (VOCs), such as aldehydes, alcohols, esters, and terpenoids, are released from plant flowers, vegetative parts or roots to attract pollinators and predators, repel herbivores, and communicate between or within plants [56, 57]. When a plant is attacked, it is able to communicate with other plants and alert them of a possible future attack [58]; thereby, the alerted plants will respond stronger once attacked [59]. For example, when molasses grass, *Melinis minutiflora*, was planted in a maize field, the herbivore damage decreased. The grass emits a compound in response to caterpillar damage to attract parasitoids, and the amount of caterpillar in a maize decreased by parasitoids, after induction of JA to release more VOCs [60, 61].

The perception of herbivory by plants involved not only mechanical injury to plant and the presence of herbivore-derived elicitors released during feeding but also the presence of microbes associated with the herbivore [62]. Microbial symbionts can influence their hosts including providing nutrition, digestion, and detoxifying toxins; insect symbionts have a role in mediating plant defenses [60]. Different microbes in insects may have species-specific effects on different host plants, specifically herbivores' microbiota are perceived by plants during herbivory and thus may alter the outcome plant responses [62].

7. Plant defenses against herbivores and fitness

Insects find and select their host plants and deal with plant defenses, as well as herbivores modify plant phenotypes. However, plants interact with multiple attackers and interact at different levels of biological organization [39].

Herbivory affects the expression of floral traits, plant-pollinator interactions, and costs-benefits to controlling reproductive systems and defense strategies. Plant-herbivore interaction promotes myriad defenses that protect plants from damage. In recent years, it has been considered whether reproductive traits and antiherbivore defenses are interdependent as a result of pollinator- and herbivore-mediated selection [63]. Floral traits are most likely to affect susceptibility to herbivores. There are pollinating herbivores, which when adult insects pollinate the plants their larvae use as host, for example, figs and fig wasps [64], the larvae feed directly on ovules and developing seeds. A diversity of floral traits influences the susceptibility of plants to herbivores; for example, taller inflorescences often result in greater herbivory, phenology also affects herbivory risk, and plants that flower early or late typically receive less damage than plants that flower during peak flowering [63].

On the other hand, inbreeding can produce individuals with reduced fitness, but inbred plants are more susceptible to herbivores than outbreds [7]. In horsenettle (*Solanum carolinense* L.), the tobacco hornworm caterpillars (*Manduca sexta* L.) preferred to feed on inbred plants, and the females oviposited more frequently on inbred plants compared to outbreds [65, 66].

Inbreeding in horsenettle causes significant reduction in the plant's induced defense responses and resistance to herbivory [67–69]. The predilection for inbred plants exhibited by insects suggests that they are gaining fitness benefits by choosing inbred host plants, regulated by insect herbivore growth, oviposition, and flight capacity. Inbred plants, serve as better host for developing insects could be that inbred plants suffer from a limited ability to unregulate genes in defense biochemical pathways. In the system plant-insect horsenettle-tobacco hornworm suggests that biochemical changes in plant inbreeding can influence in the health of animals at a higher trophic level, particularly in insect herbivores which increases survival, growth, and flight metabolism when nurtured on inbred plants [7].

8. Tolerance traits

There is another plant defense strategy: tolerance. In resistance plant synthesizes structural or chemical traits to minimize herbivore damage, while in tolerance traits reduce the negative effects or herbivore damage [1].

The traits that maintain or promote plant fitness following damage before or after infestation can confer herbivore tolerance, and they are grouped in those that alter physiological process like photosynthesis and growth, phenology, and nutrient storage [1]. In many plant species, partial defoliation leads to increased photosynthetic rate in the remaining plant tissues, but is not universal [70]. Delayed growth, flower, and fruit production following herbivore damage could promote herbivore tolerance by postponing plant development until the threat of attack has passed [71].

Roots eaten by insect herbivores exhibit extensive regrowth, in density and quantity [72]. The former might be caused by additional lignification that could increase the toughness of the roots [73].

Mechanisms involved in increased tolerance are [i] increased net photosynthetic rate after damage, [ii] high relative growth rates, [iii] increased branching or tillering after release of apical dominance, [iv] preexisting high levels of carbon storage in roots for allocation to aboveground reproduction, and [v] ability to shunt carbon stores from roots to shoots after damage. The evolution of tolerance can promote an apparently mutualistic relationship between plant and herbivore populations [70].

9. Example conifer plant defense against bark beetle

Now, we examined how different responses can be used by *Pinus* genera to limit damage causes by attack of bark beetle, one of the principal plagues that affect *Pinus* populations.

Most herbivores are insects that feed on plants in various forms, for example, they adopt different feeding strategies throughout their life cycle and can feed both external [leaf buds or flowers] and internal structures of the plant [miners, stem borers, gillnet] [74].

Unlike other herbivores such as mammals, insects commonly feed on the leaves and other parts of the mature plant typically do not cause the death of the plant; as for insects to kill the plant, they will require much time [75]. Thus, the relationship between herbivorous insects and plants is more like the host-parasite than predator-prey relationship. Plants for their part have not become passive victims of herbivorous insects as they have been able to produce special metabolites and toxic proteins, which serve as repellents or have antinutritional effects for their attackers [76]. However, herbivorous insects successfully consume plant material, overcoming the complex set of defenses of plant [74, 77, 78]. Moreover, unlike other herbivores, insects are much more specialized, because they can feed exclusively from a plant species or a limited number of them [75, 79, 80]. Therefore, it is necessary to understand the relationship between herbivorous insects and their host plants from biochemical, ecological, behavioral, physiological, and genetic aspects, including the ways in which insects can affect the abundance and distribution of plant species [75].

9.1. Herbivory and regulation of plant populations

Herbivorous insects usually cause reduced growth, fertility, and even the survival of plants; some plants can counter or overcompensate significant amounts of damage in general [75, 81]; however, the insect damage as a group causes a multiple effect and simultaneously in succession with additive effects and multiplicative on the plant fitness, which results in a significant impact on the abundance of plants, distribution, or population dynamics [82].

The role of herbivorous insects in the regulation of plant populations and dynamics of communities has been poorly documented; most studies have focused mainly on explaining the role of herbivorous insects' native as agents that limit the distribution of its plant host [75]. However, it has been possible to distinguish that the effects of herbivorous insects on plants may differ depending of the different scenarios under which the interaction takes place as in the case of herbivorous insects (bark beetles) and pines.

On the one hand, if the evolutionary success involves adaptive radiation and overtime, the species survive and expand their geographical distribution, and then pines (*Pinus* sp.) can be considered successful, because they form the largest genus of conifers in the Pinaceae family. The pine group consists of more than 100 species, many subspecies, and varieties. Although mainly distributed in temperate regions of the northern hemisphere, pines also occupy other habitats and climates [83, 84].

Moreover, the great success of the pines can be attributed to their defense strategies against herbivorous insects or parasites [85]. For its wide distribution and its prolonged generational cycles, ranging from decades to more than 4000 years such as *Pinus longaeva* [85], pines are subject to deal with a wide range of attackers at which they have developed along its evolution complex defense mechanisms [84].

The basic defense strategy of conifers including pines is both morphological structures [physical barriers] and chemical mechanisms [85]. Physical barriers are formed by static structures such as lignified cells, calcium oxalate crystals, or hard foliage; they act primarily against herbivores, ovipositors, and defoliating insects [84, 85]. The bark of the trunk on his part is of particular interest because it forms the first barrier against herbivorous insects such as bark beetles, whose evolution has specialized to kill the tree [85]. Then, conifers produce a plethora of chemical defenses where the most important are phenolic compounds and oleoresins which contain numerous terpenoids. Chemical defense mechanisms may be directed against herbivorous insects to prevent oviposition and food or affect their physiology to reduce survival or fecundity [86].

9.2. Defense and resistance strategies of conifers against bark beetles and fungal pathogens

Conifers throughout their life cycle face the challenges of a variety of organisms cycle, conifers face the challenges of a variety of organisms, the more severe are the bark beetle and fungal pathogens associated [85]. Conifer defenses against insects and pathogens that infect the trunk are classified as constitutive and induced [84, 85].

9.2.1. Constitutive defense systems

Mechanisms that produce a stable set of structural defenses (cells and resin canals), toxic chemicals such as phenolics and terpenes, and mechanical properties of the cortex (suberized layers of cells and lignified oxalate crystal calcium) are permanent. The constitutive systems are defenses with great resilience against a number of organisms trying to penetrate the cortex during the history of the tree and against common secondary invasions of opportunistic organisms. The constitutive defenses are of two basic types:

A. *Mechanical defenses*: Structural elements that provide hardness or thickness to tissues and inhibit mastication or piercing in the bark. Impregnating plant tissues with polymers such as suberin and lignins can add resistance to the mechanical properties against penetration, degradation, and ingestion/mastication by insects.

B. *Chemical defenses*: Formed by chemical compounds stored, like phenolics, terpenoids, and alkaloids, and released under attack. Antinutritive defenses include chemical, toxins, defensive proteins, enzymes, and resin deposits that can flow to repel or physically trap small organisms. These defenses are scattered in the tissues of the bark [periderm, cortex, and secondary phloem]. The constitutive strategies vary depending on the physical or chemical nature of defense and its distribution within the bark and trunk [85].

9.2.1.1. Periderm defenses

Periderm forms a permeable barrier for controlling the gas exchange in the trunk and is the first line of defense against biotic and abiotic factors. It is characterized by the presence of multiple layers of cells, most of which are dead, are also structurally and chemically different, and have lignified or suberized its walls. Cells may contain high amounts of phenolic compounds, and one or more layers have encrusted calcium oxalate crystals. These mechanical defenses (hard walls lignified, crystallization, and suberization) provide a hydrophobic barrier, combined with the chemical properties of the phenolic compounds and form a multifunctional barrier against the external environment. However, the periderm is not a continuous barrier, due to the presence of lenticels to allow gas exchange at the surface, although it is not an open system that may allow entry of invading organisms as in the case of small bark beetles (*Pityogenes chalcographus*) in *Picea abies* [87, 88].

9.2.1.2. Cortex defenses

The cortex is formed during the early development of the stem, so it is an important general barrier, especially during the early development of the stem. It remains alive for several years during the secondary growth and contains high amounts of phenolic compounds within vacuoles of cortical parenchyma; in many Pinaceae, the cortex has axial duct resins, which participate in defense, although its function is replaced by the secondary phloem [87].

9.2.1.3. Secondary phloem defenses

The secondary phloem is the most important site of constitutive defense mechanisms of conifers and is made up of phenolic bodies, sclerenchyma, and calcium oxalate crystals; the relative amount of these components varies considerably between species [89]. A fourth constitutive strategy of defense in certain taxa as Pinaceae is the production of resin structures comprising radial ducts extended from xylem, axial ducts, blisters, and resin cells. The amount and combination of each of these components define defense strategies. In the secondary phloem, there are specialized structures, such as phenolic bodies, sclerenchyma, calcium oxalate crystals, and resins.

The phenolic bodies are parenchymal cells of the axial phloem, also called polyphenolic parenchymal cells [PP cells], specializing in the synthesis and storage of phenolic compounds [89, 90], making nonedible tissues or antifungal capacity [85, 91]. Different species produce different phenolic compounds depending on the type of organisms that commonly attack, so that the relative resistance to pathogens may be due in part to the type of phenolic compounds they produce [92, 93].

Moreover, the PP cells are responsible for responses of induced defense and, even when they have thickened walls, allow the exchange of axial and tangential information and signaling for defense because they contain lots of plasmodesmata. The PP cells represent a very dynamic component in defense strategies in conifers and are most abundant in the secondary phloem [94]. Another important feature of the PP cells located along the radial ducts parenchyma is that they are an important site that stores starch and/or lipids [94], which are considered the target for bark beetles and fungi; however, the presence of phenolic compound constituent allows cells to protect themselves and prevent the penetration of fungi into the area of the cambium. In any case, the layers of PP cells form a sieve maintaining the physical and chemical resistance to prevent penetration into the cortex [95].

Another important tissue with mechanical function is the sclerenchyma, which is common in the bark of conifers; quantity and type vary among taxa. It consists in cells with thickened lignified secondary wall, which are known as "stone cells" because they are high hardness cells or sclereids, so they can serve as structural element and mechanical defense. This organization is massive and irregular in many Pinaceae or organized form rows as in the case of Taxaceae [96]. Their physical strength can detain predation or perforation of the bark by insects forming a screen of dead cells that progressively collapse under pressure of new layers of inner cells [90, 97–99].

The crystals of calcium oxalate formed are stored intracellularly in the secondary phloem of conifers, particularly in Pinaceae, and represents a defense mechanism because the physical nature of the crystals and their relative abundance could imply a role in deterring penetration bark or chewing by herbivores. However, being chemically inert, it is unlikely to have any effect on fungal attack [85].

One of the common deposits in plants is crystals of calcium oxalate [85, 100–103], and its role in most of them is the regulation of calcium [104–106]; however, also they have secondary functions of defense [85, 107]. In Pinaceae the calcium oxalate crystals embedded within the phenolic bodies in PP cells vacuoles present typically form scattered axial lines crystallized cells. The combination of several layers of fibers and dense encrustation with crystals can provide a powerful defense against bark beetles [85].

One of the main constitutive defenses is resins, particularly for Pinaceae. The resin production and storage structures for this include radial resin ducts, axial ducts or channels, blisters, and resin cells. Ducts and blisters have a coating epithelial cell enriched by plastids that synthesize terpenoid resins and secreted into the extracellular lumen, which is accumulated under pressure. After injuries are caused by damage from invading organism, the pressed resin is released and may expel the invading organism from the bark and catch it thanks to its sticky consistency or kill it because of its toxic nature. Volatile resin components evaporate and non-volatile crystallize to sterilize and seal the damaged region effectively. It has been shown that the resin is an effective defense against insect bark borers [108].

9.2.1.4. Secondary xylem defenses

Secondary xylem is a general system of defense in trunk, which is involved in the synthesis and storage of resin and phenolic compounds and other secondary products such as lignins [109], and provides a defense against wood-rotting fungi and other organisms. The constituent axial ducts of resin found in the xylem of some conifers can contribute to resin flow when connected to the radial ducts that traverse the xylem and phloem [110].

9.2.2. Induced defense systems: second level of defense

Induced defense system or responses due to herbivore attack involves the synthesis "de novo" or activation of a wide range of chemical defenses, including terpenoids, phenolic compounds, PR proteins, reactive oxygen species, and enzymes. The induced defense system can act against a current infection presenting a hypersensitive response and local resistance or against future infections or attacks by bark beetles generating responses with acquired resistance [85, 111].

A. *Induced structural defenses*: Structural defenses in bark are important, because they improve the overall defense capability of the plant; these are diverse and include structural changes and synthesis of chemical and biochemical agents. They are a combination of responses apparently targeting specific organisms, including the general increase in hypersensitivity responses, aimed at limiting the spread of detected damage and isolating the invading organism, repairing damaged tissues, and limiting the attack or later invasion of opportunistic organisms. In addition, long term results in acquired resistance [85, 111, 112]. Among these structural defenses are hypersensitive response, callus tissue formation, and scarring in the periderm.

9.2.2.1. Hypersensitive response

Damage produces a hypersensitive response in the plant, which quickly stops invading organisms sacrificing a small piece of tissue [112]. The hypersensitive response occurs locally at the site of infection or attack, producing reactive oxygen species causing rapid cell death, which tries to stop organisms such as pathogenic fungi, bacteria, and virus killing only the damaged plant tissue that has been attacked [87, 111].

9.2.2.2. Callus tissue formation

A more generalized response in the case of wounds in plants is the formation of callus tissue that can subsequently lignify, suberize, or impregnate phenolic compounds to provide a barrier, part of the wound periderm. This reaction provides protection against new intrusions and blocking an organism such as a fungal pathogen. The callus can also repair damaged tissues so that its functions can be restored [87].

9.2.2.3. Scarring in the periderm

Periderm scars are produced around damaged regions of the cortex, which cause activation of the PP cells of the secondary phloem, which begin to divide to form new tissue. Periderm scar acts as a wall that essentially isolates the damaged area preventing the supply of nutrients to the wound area, which eventually dies if not already dead by the attack of an invading organism. These scars also have permanent effects of tissue repair and generally are formed within the limits of induced injuries by bark beetles or fungal attacks in the trunks of conifers or well around any damaged tissue [112].

B. *Induced chemical defenses*: While the constituent chemical defenses are generally nonselective for pest species, induced chemical defenses can be broad-spectrum and specific components. Chemical defenses are extremely diverse and therefore cover a wide range of pests. Nonprotein chemicals, such as products of the phenylpropanoid routes (phenolic) and isoprenoids (terpenoids resin) products, as well as alkaloids can have potent effects on invading organisms.

These compounds are produced more rapidly than protein-based defense because the path usually exists in tissues and only requires activation. However, some of the biochemical pathways are created "de novo" in the tissues [90, 96].

Another advantage of these chemical defenses is often effective against a wide range of organisms and thus may delay an attack, while recognition mechanisms come into play to identify the organism and then activate specific defenses against herbivore [77]. Among chemical compound induced by herbivores in conifers are phenols, resin terpenoids, and proteins.

9.2.2.4. *Phenolic compounds*

Phenolic compounds are abundant in the bark of conifers [113–115], mainly in the PP cells. Both phenolic compounds and tannins act as antifungal agents and block hydrolytic enzymes secreted by invading organisms, thereby inhibiting its progress in tissues [116–118]. By joining amino acids and proteins disturbed by plant tissues, phenolics and tannins reduce the nutritional value for attackers while coupling to digestive enzymes in the intestine decreases the ability to digest plant tissues. The wounds of the plant or invading organisms in the cortex activate PP cells, including cell expansion and accumulation of a higher amount of phenolic compounds [95, 119, 120]. Generally, the induced phenolic compounds are more toxic or more specific to an invading organism than the constituent phenols, whereby the conversion of polyphenolic compounds to soluble phenolic compounds during an attack adds to the defense capacity; evidence of this is the reduction of polyphenols in vacuoles of intact cells PP near the region of attack [121].

9.2.2.5. *Resin terpenoids*

Resin terpenoid production is induced by the attack of organisms. During and after attack, the resin flow in the wound can be quite extensive, especially in the Pinaceae. Part of this resin is stored in the structures that produce, while the constituent ducts can be activated to produce resin [89, 122, 123].

Within the first 2–3 weeks of the attack, the new resin ducts are induced to produce, being considered as traumatic resin ducts [124–127], and the resin forming these ducts can be different from the constitutive resin [103, 128, 129]. In Pinaceae and some other groups of conifers, traumatic ducts are formed in the xylem [130] and interconnected with the radial ducts phloem [131]. However, some species of conifers are induced to form more traumatic ducts in the phloem and the xylem [89]. Regardless of their origin, the end result of the development of traumatic resin ducts is to increase the formation and accumulation of resin and increase its flow [128, 129]. The increased flow helps to kill or expel the invaders and to seal the wound and resin-soaked regions of the bark and wood making them more resistant to microbial activity. Furthermore, it has been found that traumatic ducts can confer acquired resistance to subsequent attacks [131, 132] and the resin in traumatic ducts may be more toxic through changes of terpenoids or addition of phenolic compounds [133].

9.2.2.6. *Proteins*

Chemical defenses of the trees based on proteins include enzymes such as chitinases and glucanases that may degrade components of invading organisms and toxic proteins such as porins, lectins, and enzyme inhibitors such as proteinases and amylase. Inhibiting enzymes interfere with the ability of the invading organism to use resources from invaded tissue. Other induced enzymes such as peroxidases and laccases can do more resistant cell walls through crisscrossed reactions or promotion of lignification or well included affecting invader organism. The protein-based defenses can be highly specific for certain organisms. For example, in Norway spruce, there are chitinases as a large family of proteins, but only a small subset of them can be regulated during the attack by a specific pathogenic fungus [133, 134], and it is presumed that these are effective against the wall cell of this organism. In general, chemical defenses induced mechanical follow a pattern similar to the induced structural defense, such as overlapping of multiple strategies. The production of a toxic cocktail with various chemical components maximizes the potential to stop or destroy an aggressive or virulent invading organism, in contrast to a more conservative production of one or few directed defenses.

9.2.3. Remark defense system importance in conifers

Multiple overlaying of structures and defense systems provides an efficient barrier against a wide range of possible attacks of organisms. However, conifers remain susceptible to certain organisms that have evolved strategies to overcome the defenses or avoid them. Nevertheless, the remarkable longevity of various species of conifers is a testament to the success of their defense strategies [87].

The first line of defense of the plant is given by a mechanical resistance to attack, due to the hardened cells either by thickening the walls or storing different compounds like calcium oxalate crystals that are joined to form a screen of high hardness. This first defense system is effective against most of the organisms that can attack the tree; however, bark beetles usually manage to overcome this barrier, bringing with them pathogenic fungi.

After that penetrate the bark beetles, thanks to its powerful masticatory apparatus, tree active chemical defense mechanisms, in which the phenolic bodies, resin and some proteins may be directed mainly beetles as organisms that are directly attacking the tree; however, these compounds also have an effect on fungi. Another unspecific compounds may function to attack bark beetles as in the case of some proteins and calcium oxalate crystals during the attack the hypersensitive response is activated, the formation of callous bodies and interaction with proteins and enzymes which are directed primarily by fungal attack. Also, answers that could be used for both bark beetles and fungi, as in the case of periderm scars, phenolic compounds, and terpenoids, can be triggered. But nevertheless, together, the beetle and the fungus can gradually block the tree's defenses, weakening to lead to death.

10. Conclusion

Plants have been developing a plethora of defense traits: chemical direct and indirect, mechanic, and uses interactions. All defense mechanisms aim survival with high photosynthetically rates, population maintenance, and fitness for plant.

Author details

Hermilo Sánchez-Sánchez[1] and Alina Morquecho-Contreras[2]*

*Address all correspondence to: alimorcon@gmail.com

1 Centro de Investigación en Recursos Bióticos, Universidad Autónoma del Estado de México, Toluca, Mexico

2 Facultad de Ciencias Agrícolas, Universidad Autónoma del Estado de México, Toluca, Mexico

References

[1] Mitchell C, Brennan RM, Graham J, Karley AJ. Plant defense against herbivorous pests: exploiting resistance and tolerance traits for sustainable crop protection. Front Plant Sci. 2016; **7**: 1132. DOI: 10.3389/fpls.2016.01132.

[2] Wang M, Bezemer TM, van der Putten W, Biere A. Effects of the timing of herbivory on plant defense induction and insect performance in *Ribwort plantain* (*Plantago lanceolata* L.) depend on plant mycorrhizal status. J Chem Ecol. 2015; **41**: 1006–1017. DOI: 10.1007/s10886-15-0644-0.

[3] Sato Y, Kudoh H. Associational effects against a leaf beetle mediate a minority advantage in defense and growth between hairy and glabrous plants. Evol Ecol. 2016; **30**:137–154. DOI 10.1007/s10682-015-9809-0.

[4] Bixenmann RJ, Coley PD, Weinhold A, Kursar TA. High herbivore pressure favor constitutive over induced defense. Ecol Evol. 2016; **6**. DOI: 10.1002/ece3.2208.

[5] Turcotte MM, Davies TJ, Thomsen CJM, Johnson MTJ. Macroecological and macroevolutionary patterns of leaf herbivory across vascular plants. Proc R Soc B. 2014; **281**:20140555. DOI: 10.1098/rspb.2014.0555.

[6] Preisser EL, Gibson SE, Adler LS, Lewis EE. Underground herbivory and the costs of constitutive defense in tobacco. Acta Oecol. 2007; **31**: 210–215. DOI:10.1016/j.actao.2006.09.004.

[7] Portman SL, Kariyat RR, Johnston MA, Sthephenson AG, Marden JH. Inbreeding compromises host plant defense gene expression and improves herbivore survival. Plant Signal Behav. 2015; **10**: 5, e998548, DOI: 10.1080/15592324-2014.998548.

[8] War AR, Paulraj MG, Ahamad T, Buhroo AA, Hussain B, Ignacimuthu S, Sharma HC. Mechanisms of plant defense against insect herbivores. Plant Signal Behav. 2012; **7**: 1306–1320. DOI: 10.4161/psb.21663.

[9] Brandoly M, Hilker M, Steppuhn A. Oviposition by *Spodoptera exigua* on *Nicotiana attenuata* primes induced plant defense against larval herbivory. Plant J. 2015; **83**: 661–672. DOI: 10.111/tpj.12918.

[10] Scholz SS, Heyer M, Vadassery J, Mithöfer A. A role for calmodulin-like proteins in herbivore defense pathways in plants. J Endocytobiosis Cell Res. 2016; **27**: 1–12.

[11] Fürstenberg-Hägg J, Zagrobelny M, Bak S. Plant defense against insect herbivores. Int J Mol Sci. 2013; **14**:10242–10297.

[12] Bonaventure G, Van Doorn A, Baldwin IT. Herbivore-associated elicitors: FAC signaling and metabolism. Trends Plant Sci. 2011; **16**: 294–299.

[13] Huffaker A, Pearce G, Veyrat N, Erb M, Turlings TCJ, Sartor R, Shen Z, Briggs SP, Vaughan MM, Alborn HT, Teal PEA, Schmeiz EA. Plant elicitor peptides are conserved signals regulating direct and indirect antiherbivore defense. Proc Natl Acad Sci U S A. 2013; **110**: 5707–5712. DOI: 10.1073/pnas.1214668110.

[14] Maffei ME, Arimura G, Mithöfer A. Natural elicitors, effectors and modulators of plant responses. Nat Prod Rep. 2012; **29**: 1288–1303.

[15] Maffei ME, Mithöfer A, Boland W. Before gene expression: early events in plant-insect interaction. Trends Plant Sci. 2007; **12**: 310–316.

[16] Pieterse CMJ, Van der Does D, Zamioudis C, Leon-Reyes A, Van Wees SC. Hormonal modulation of plant immunity. Annu Rev Cell Dev Biol. 2012; **28**: 489–521.

[17] Rosenthal GA, Berenbaum MR. Herbivores: their interactions with secondary plant metabolites. 2nd ed. Academic Press. 452 p. ISBN: 9780125971836.

[18] Vandenborre G, Smagghe G, Van Damme EJ. Plant lectins as defense proteins against phytophagous insects. Phytochemistry. 2011; **72**: 1538–1550.

[19] Howe GA, Jander G. Plant immunity to insect herbivores. Annu Rev Plant Biol. 2008; **59**: 41–66.

[20] Petterson DS, Harris DJ, Allen DG. Alkaloids. In Toxic substances in crop plants; D'Mello JPF, Duffus CM, Duffus JH, ed.; The Royal Society of Chemistry, Cambridge, UK. 1991; pp. 148–179.

[21] Ziegler J, Facchini PJ. Alkaloid biosynthesis: metabolism and trafficking. Annu Rev Plant Biol. 2008; **59**: 736–769.

[22] Cheeke PR. Toxicants of plant origin. 1989, Volume IV, Phenolics; Boca Raton CRC Press, Boca Raton, FL, USA.

[23] Nuessly GS, Scully BT, Hentz MG, Beiriger R, Snook ME, Widstrom NW. Resistance to *Spodoptera frugiperda* (lepidoptera: Noctuidae) and *Euxesta stigmatias* (diptera: Ulidiidae) in sweet corn derived from exogenous and endogenous genetic systems. J Econ Entomol. 2007; **100**: 1887–1895.

[24] Luczynski A, Isman MB, Rawirth DA. Strawberry foliar phenolics and their relationship to development of the two-spotted spider mite. J Econ Entomol. 1990; **83**: 557–563.

[25] Ruuhola T, Tikkanen O, Tahvanainen O. Differences in host use efficiency of larvae of a generalist moth, *Operophtera brumata* on three chemically divergent salix species. J Chem Ecol. 2001; **27**: 1595–1615. 246.

[26] Aharoni A, Jongsma MA, Bouwmeester HJ. Volatile science? Metabolic engineering of terpenoids in plants. Trends Plant Sci. 2005; **10**: 594–602.

[27] Cherrett JM. Some factors involved in the selection of vegetable substrate by *Atta cephalotes* (L.) (hymenoptera: Formicidae) in tropical rain forest. J Anim Ecol. 1972; **41**: 647–660.

[28] Hummelbrunner LA, Isman MB. Acute, sublethal, antifeedant, and synergistic effects of monoterpenoid essential oil compounds on the tobacco cutworm, *Spodoptera litura* (Lep., Noctuidae]). J Agric Food Chem. 2001; **49**: 715–720.

[29] Scholz SS, Reichelt M, Mekonnen DW, Ludewing F, Mithöfer A. Insect herbivory-elicited GABA accumulation in plants is a wound-induced, direct, systemic, and jasmonate-independent defense response. Front Plant Sci. 2015; **6**: 1128. DOI: 10.3389/fpls2015.01128.

[30] Lucas PW, Turner IM, Dominy NJ, Yamashita N. Mechanical defences to herbivory. Ann Bot. 2000; **86**: 913–920. DOI: 10.1006/anbo.2000.1261.

[31] Agrawal A, Karban R. Why induced defenses may be favored over constitutive. pp 45–65 in Tollrian R, Harvell CD. The ecology and evolution of inducible defenses. 1999. Princeton Univ. Press, Princeton. DOI:10.1234/12345678.

[32] Aziz M, Nadipalli RK, Xie X, Sun Y, Surowiec K, Zhang JL Pare PW. Augmenting sulfur metabolism and herbivore defense in *Arabidopsis* by bacterial volatile signal. Front Plant Sci. 2016; **7**: 458. DOI: 10.3389/fpls.2016.00458.

[33] Kachroo A, Kachroo P. Fatty acid-derived signals in plant defense. Annu Rev Phytopatol. 2009; **47**: 153–176. DOI: 10.1146/annurev-phyto-080508-081820.

[34] Hettenhausen C, Hernrich M, Baldwin I, Wu J. Fatty acid-amino acid conjugates are essential for systemic activation of salicylic acid-induced protein kinase and accumulation of jasmonic acid in *Nicotiana attenuata*. BMC Plant Biol. 2014; **14**:326. DOI: 10.1186/s12870-014-0326-z.

[35] Tsao R, Marvin CH, Broadbent AB, Friesen M, Allen WR, McGarvey BD. Evidence for an isobutylamide associated with host-plant resistance to western flower thrips, *Frankliniella occidentalis*, in chrysanthemum. J Chem Ecol. 2005; **31**: 103–110.

[36] Morquecho-Contreras A, López-Bucio J. Cannabinoid-like signaling and other new developmental pathways in plants. Int J Plant Dev Biol. 2007; **1**: 34–41.

[37] Lait CG, Alborn HT, Teal PE, Tumlison JH. Rapid biosynthesis of N-linolenoyl-L-glutamine, an elicitor of plant volatiles, by membrane associated enzyme(s) in Manduca sexta. Proc Natl Acad Sci USA. 2003; **100**: 7027–7032.

[38] Thines B, Katsir L, Melotto M, Niu Y, Mandaokar A. JAZ repressor proteins are targets of the SCF (COI1) complex during jasmonate signalling. Nature. 2007; **448**: 661–665.

[39] Stam JM, Kroes A, Li Y, Gols R, van Loon JJA, Poelman EH, Dicke M. Plant interactions with multiple insect herbivores: from community to genes. Annu Rev Plant Biol. 2014; **65**: 689–713. DOI: 10.1146/annurev-arplant-050213-035937.

[40] Lorenzo O, Chico JM, Sa JJ. JASMONATE-INSENSITIVE1 encodes a MYC transcription factor essential to discriminate between different jasmonate-regulated defense responses in Arabidopsis. Plant Cell. 2004; **16**: 1938–1950.

[41] Dempsey DA, Vlot AC, Wildermuth MC, Klessig DF. Salicylic acid biosynthesis and metabolism. Arabidopsis Book. 2001; **9**: e0156.

[42] Durrant WE, Dong X. Systemic acquired resistance. Annu Rev Phytopathol. 2004; **42**: 185–209.

[43] Reymond P, Bodenhausen N, Van Poecke RMP, Krishanamurthy V, Dicke M, Farmer EE. A conserved transcript pattern in response to a specialist and generalist herbivore. Plant Cell. 2004; **16**: 3132–3147.

[44] Yamawo A, Suzuki N, Tagawa J, Hada Y. Leaf ageing promotes the shift in defence tactics in *Mallotus japonicus* from direct to indirect defense. J Ecol. 2012; **100**: 802–809. DOI: 10.1111/j.1365-2745.2011.01934.x.

[45] Hartley SE, Fitt RN, McLarnon EL, Wade RN. Defending the leaf surface: intra and interspecific differences in silicon deposition in grasses in response to damage and silicon supply. Front Plant Sci. 2015; http://dx.doi.org/10.3389/fpls.2015.00035.

[46] Schaller A, Howe GA. Direct defenses in plants and their induction by wounding and insect herbivores.

[47] Schoonhoven LM, van Loon JJA, Dicke M. Insect-plant biology. 2005. Oxford University Press, Oxford, UK.

[48] Agrawal AA, Konno K. Latex: a model for understanding mechanisms, ecology, and evolution of plant defense against herbivory. Annu Rev Ecol Syst. 2009; **40**: 311–331.

[49] Dussourd DE, Denno RF. Deactivation of plant defense: correspondence between insect behavior and secretory canal architecture. Ecology. 1991; **72**: 1383–1396.

[50] Phillips MA, Croteau RB. Resin-based defenses in conifers. Trends Plant Sci. 1999; **4**:184–190.

[51] Dicke M, Sabelis MW. How plants obtain predatory mites as bodyguards. Neth J Zool. 1988; **38**: 149–165.

[52] Wäckers FL, Zuber D, Wunderlin R, Keller F. The effect of herbivory on temporal and spatial dynamics of foliar nectar production in cotton and castor. Ann Bot. 2001; **87**: 365–370.

[53] Heil M. Indirect defence via tritrophic interactions. New Phytol. 2007; **178**: 41–61.

[54] Ballorn DJ, Kay J, Kautz S. Quantitative effects of leaf area removal on indirect defense of Lima bean (*Phaseolus lunatus*) in nature. J Chem Ecol. 2014; **40**: 204. DOI: 10.1007/s10886-014-0392-6.

[55] Kim TN, Underwood N. Plant neighborhood effects on herbivory: damage is both density and frequency dependent. Ecology. 2015; **96**: 1431–1437. DOI: 10.1890/14-1097.1.sm.

[56] Kessler A, Baldwin IT. Defensive function of herbivore-induced plant volatile emissions in nature. Science. 2001; **291**: 2141–2144.

[57] Dudareva N, Negre F, Nagegowda DA, Orlova, I. Plant volatiles: recent advances and future perspectives. Crit Rev Plant Sci. 2006; **25**: 417–440.

[58] Engelberth J, Alborn HT, Schmelz EA, Tumlinson JH. Airborne signals prime plants against insect herbivore attack. Proc Natl Acad Sci USA. 2004; **101**: 1781–1785.

[59] Zimmerli L, Jakab G, Métraux JP, Mauch-Mani B. Potentiation of pathogen-specific defense mechanisms in Arabidopsis by β-aminobutyric acid. Proc Natl Acad Sci U S A. 2000; **97**: 12920–12925.

[60] Khan ZR, AmpongNyarko K, Chiliswa P, Hassanali A, Kimani S, Lwande W, Overholt WA, Pickett JA, Smart LE, Wadhams LJ. Intercropping increases parasitism of pests. Nature. 1997; **388**: 631–632.

[61] Bernasconi Ockroy ML, Turlings TCJ, Edwards PJ, Fritzsche-Hoballah ME, Ambrosetti L, Bassetti P, Dorn S. Response of natural populations of predators and parasitoids to artificially induced volatile emissions in maize plants (Zea mays L.). Agric. For. Entomol. 2001; 3:201-209.

[62] Wang J, Chung SH, Peiffer M, Rosa C, Hoover K, Zeng R, Felton W. Herbivore oral secreted bacteria trigger distinct defense responses in preferred and non-preferred host plants. J Chem Ecol. 2016; **42**: 463. DOI: 10.1007/s10886-016-0712-0.

[63] Johnson MT, Campbell SA, Barrett SCH. Evolutionary interactions between plant reproduction and defense against herbivores. Annu Rev Ecol Evol Syst. 2015; **46**: 191–213. DOI: 10.1146/annurev-ecolsys-112414-054215.

[64] Cook JM, Rasplus JY. Mutualists with attitude: coevolving fig wasps and figs. Trends Ecol Evol. 2003; **18**: 241–248.

[65] Delphia CM, Stephenson AG, De Moraes, CM, Mescher M. Inbreeding in horsenettle influences host-plant quality and resistance to herbivory. Ecol Entomol. 2009; **34**: 513–519. DOI:10.1111/j.1365-2311.2009.01097.x.

[66] Kariyat RR, Mauck KE, Balogh CM, Stephenson AG, Mescher MC, De Moraes CM. Inbreeding in horsenettle (*Solanum carolinense*) alters night-time volatile emissions that guide oviposition by *Manduca sexta* moths. Proc R Soc Lond B Biol Sci. 2013; **280**: 1757. DOI: 10.1098/rspb.2013.002.

[67] Stephenson AG, Leyshon B, Travers SE, Hayes CN, Winsor JA. Interrelationships among inbreeding, herbivory, and disease on reproduction in a wild gourd. Ecology. 2004; **85**: 3023–3034. DOI: 10.1890/04-0005.

[68] Bello-Bedoy R, Nunez-Farfan J. The effect of inbreeding on defense against multiple enemies in *Datura stramonium*. J Evolut Biol. 2011; **24**: 518–530. DOI:10.1111/j.1420-9101.2010.02185.x.

[69] Kariyat RR, Scanlon SR, Moraski RP, Stephenson AG, Mescher MC, De Moraes CM. Plant inbreeding and prior herbivory influence the attraction of caterpillars (*Manduca sexta*) to odors of the host plant *Solanum carolinense* (Solanaceae). Am J Bot. 2014; **101**. PMID:24509799. DOI: 10.3732/ajb.1300295.

[70] Strauss SY, Agrawal AA. The ecology and evolution of plant tolerance to herbivory. Trends Ecol Evol. 1999; **14**: 179–185. DOI: 10.1016/s0169-5347(98)01576-6.

[71] Tiffin P. Mechanisms of tolerance to herbivore damage: what do we know? Evol Ecol. 2000; **14**: 523–536. DOI: 10.1023/A:1010881317261.

[72] Quinn MA, Hall MH. Compensatory response of a legume root-nodule system to nodule herbivory by *Sitona hispidulus*. Entomol Exp Appl. 1992; **64**: 167–176.

[73] Johnson SN, Hallett PD, Gillespie TL, Halpin C. Below-ground herbivory and root toughness: a potential model system using lignin-modified tobacco. Physiol Entomol. 2010; **35**: 186–191.

[74] Houchuli FD. Insect herbivory and ontogeny: how do growth and development influence feeding behaviour, morphology and host use?. Austral Ecol. 2001; **26**: 563–570.

[75] Herrera CM, Pellmyr O. Plant-animal interactions. An evolutionary approach. 2002. Blackwell Publishing, USA, UK and Australia. 313p.

[76] Zhu-Salzman K, Luthe DS, and Felton GW. Arthropod-inducible proteins: broad spectrum defenses against multiple herbivores. Plant Physiol. 2008; **146**: 852–858.

[77] Bernays EA. Evolution of feeding behavior in insect herbivores: success seen as different ways to eat without being eaten. Bioscience. 1998; **48**: 35–44.

[78] Schoonhoven LM, Jermy T, van Loon JJA. Insect-plant biology: from physiology to evolution. 1998. Chapman &Hall, UK.

[79] Wilcox JA. Leaf beetle host plants in North America. Marlton. 1979. World Natural History Publications, USA.

[80] Janzen, DH. Ecological characterization of a Costa Rican dry forest caterpillar fauna. Biotropica. 1988; **20**: 120–135.

[81] Bigger DS, and Marvier MA. How different would a world without herbivory be? A search for generality in ecology. Integr Biol. 1998; **1**: 351–360.

[82] Maron JL, Cron E. Herbivory: effects on plant abundance, distribution and population growth. Proc R Soc B. 2006; **273**: 2572–2584.

[83] Richardson DM, Rundel PW. Ecology and biogeography of pinus: an introduction. In Ecology and biogeography of pinus. Richardson DM, ed.; Cambridge University Press. England. 1998; pp. 3–46.

[84] Mumm R, Hilker M. Direct and indirect chemical defence of pine against folivorous insects. Trends Plant Sci. 2006; **11**: 352–358.

[85] Franceschi VR, Nakata PA. Calcium oxalate in plants: formation and function. Annu Rev Plant Biol. 2005; **56**: 41–71.

[86] Larsson S. Influence of plant quality on pine sawfly population dynamics. Oikos. 2000; **89**: 440–450.

[87] Franceschi VR. Krokene P, Christiansen E. Krekling T. Anatomical and chemical defenses of conifer bark against bark beetles and other pests. New Phytol. 2005; **167**: 353–376.

[88] Rosner S, Führer E. The significance of lenticels for successful *Pityogenes chalcographus* (Coleoptera: Scolytidae) invasion of Norway spruce trees (*Picea abies* (Pinaceae)). Trees. 2002; **16**: 497–503.

[89] Hudgins JW, Franceschi VR. Methyl jasmonate-induced ethylene production is responsible for conifer phloem defense responses and reprogramming of stem cambial zone for traumatic resin duct formation. Plant Physiol. 2004; **13**: 2134–2149.

[90] Hudgins JW, Christiansen E, Franceschi VR. Methyl jasmonate induces changes mimicking anatomical and chemical defenses in diverse members of the Pinaceae. Tree Physiol. 2003; **23**: 361–371.

[91] Beckman CH. Phenolic-storing cells: keys to programmed cell death and periderm formation in wilt disease resistance and in general defence responses in plants? Physiol Mol Plant Pathol. 2000; **57**: 101–110.

[92] Brignolas F, Lacroix B, Lieutier F, Sauvard D, Drouet A, Claudot AC, Yart A, Berryman AA, Christiansen E. Induced responses in phenolic metabolism in two Norway spruce clones after wounding and inoculation with *Ophiostoma polonicum*, a bark beetle-associated fungus. Plant Physiol. 1995; **109**: 821–827.

[93] Bonello P, Blodgett JT. *Pinus nigra-Sphaeropsis sapinea* as a model pathosystem to investigate local and systemic effects of fungal infection of pines. Physiol Mol Plant Pathol. 2003; **63**: 249–261.

[94] Krekling T, Franceschi VR, Berryman AA, Christiansen E. The structure and development of polyphenolic parenchyma cells in Norway spruce (*Picea abies*) bark. Flora. 2000; **195**: 354–369.

[95] Franceschi VR, Krokene P, Krekling T, Christiansen E. Phloem parenchyma cells are involved in local and distant defense responses to fungal inoculation or bark-beetle attack in Norway spruce (Pinaceae). Am J Bot. 2000; **87**: 314–326.

[96] Hudgins JW, Christiansen E, Franceschi V. Induction of anatomically based defense responses in stems of diverse conifers by methyl jasmonate: a phylogenetic perspective. Tree Physiol. 2004; **24**: 251–264.

[97] Wainhouse D, Cross DJ, Howell RS. The role of lignin as a defense against the spruce bark beetle *Dendroctonus micans*: effect on larvae and adults. Oecologia. 1990; **85**: 257–265.

[98] Wainhouse D, Rose DR, Pearce J. The influence of preformed defenses on the dynamic wound response in spruce bark. Funct Ecol. 1997; **11**: 564–572.

[99] Wainhouse D, Ashburner R, Ward E, Boswell R. The effect of lignin and bark wounding on susceptibility of spruce to Dendroctonus micans. J Chem Ecol. 1998; **24**: 1551–1561.

[100] Arnott HJ, Pautard FGE. Calcification in plants. In Biological calcification: cellular, molecular aspects. Schraer H, ed.; Appleton-Century-Crofts, New York, USA. 1970. 375–446.

[101] Franceschi VR, Horner HT Jr. Calcium oxalate crystals in plants. Bot Rev. 1980; **46**: 361–427.

[102] Webb MA. Cell-mediated crystallization of calcium oxalate in plants. Plant Cell. 1999; **11**: 751–761.

[103] Nault JR, Alfaro RI. Changes in cortical and wood terpenes in Sitka spruce in response to wounding. Can J For Res. 2001; **31**: 1561–1568.

[104] Franceschi VR. Calcium oxalate formation is a rapid and reversible process in *Lemna minor* L. Protoplasma. 1989; **148**: 130–137.

[105] Volk GM, Lynch-Holm VJ, Kostman TA, Goss LJ, Franceschi VR. The role of druse and raphide calcium oxalate crystals in tissue calcium regulation in *Pistia stratiotes* leaves. Plant Biol. 2002; **4**: 34–45.

[106] Mazen AMA, Zhang D, Franceschi VR. Calcium oxalate formation in *Lemna minor* L. physiological and ultrastructural aspects of high capacity calcium sequestration. New Phytol. 2004; **161**: 435–448.

[107] Franceschi VR. Calcium oxalate in plants. Trends Plant Sci. 2001; **6**: 331.

[108] Alfaro RI, He FL, Tomlin E, Kiss G. White spruce resistance to white pine weevil related to bark resin canal density. Can J Bot. 1997; **75**: 568–573.

[109] Kwon M, Davin LB, Lewis NG. In situ hybridization and immunolocalization of lignan reductases in woody tissues: implications for heartwood formation and other forms of vascular tissue preservation. Phytochemistry. 2001; **57**: 899–914.

[110] Wu H, Hu Z. Comparative anatomy of resin ducts of the Pinaceae. Trees. 1997; **11**: 135–143.

[111] Bleiker KP, Uzunovic A. Fast- and slow-growing subalpine fir produce lesions of different sizes in response to inoculation with a blue-stain fungus associated with *Dryocoetes confusus* (Coleoptera: Scolytidae). Can J Bot. 2004; **82**: 735–741.

[112] Christiansen E, Kucera B. Resin pockets in Norway spruce wood are not caused by the bark beetle *Ips typographus*. Rapport Fra Skogforskningen (Supplement). 1999; **12**: 1–9.

[113] Pan HF, Lundgren LN. Phenolic extractives from root bark of *Picea abies*. Phytochemistry. 1995; **39**: 1423–1428.

[114] Pan HF, Lundgren LN. Phenolics from inner bark of *Pinus sylvestris*. Phytochemistry. 1996; **42**: 1185–1189.

[115] Viiri H, Annila E, Kitunen V, Niemela P. Induced responses in stilbenes and terpenes in fertilized Norway spruce after inoculation with blue-stain fungus, *Ceratocystis polonica*. Trees Struct Funct. 2001; **15**: 112–122.

[116] Hunter RE. Inactivation of pectic enzymes by polyphenols in cotton seedlings of different ages infected with *Rhizoctonia solani*. Physiol Plant Pathol. 1974; **4**: 151–159.

[117] Nicholson RL, Hammerschmidt R. Phenolic compounds and their role in disease resistance. Annu Rev Phytopathol. 1992; **16**: 369–389.

[118] Appel HM. Phenolics in ecological interactions: the importance of oxidation. J Ecol Chem. 1993; **19**: 1521–1551.

[119] Klepzig KD, Kruger EL, Smalley EB, Raffa KF. Effects of biotic and abiotic stress on induced accumulation of terpenes and phenolics in red pines inoculated with bark beetle-vectored fungus. J Chem Ecol. 1995; **21**: 601–626.

[120] Kusumoto D, Suzuki K. Spatial distribution and time-course of polyphenol accumulation as a defense response induced by wounding in the phloem of *Chamaecyparis obtuse*. New Phytol. 2003; **159**: 167–173.

[121] Franceschi VR, Krekling T, Berryman AA, Christiansen E. Specialized phloem parenchyma cells in Norway spruce (Pinaceae) bark are an important site of defense reactions. Am J Bot. 1998; **85**: 601–615.

[122] Ruel JJ, Ayres MP, Lorio PL Jr. Loblolly pine responds to mechanical wounding with increased resin flow. Can J For Res. 1998; **28**: 596–602.

[123] Lombardero MJ, Ayres MP, Lorio PLJ, Ruel JJ. Environmental effects on constitutive and inducible resin defences of *Pinus taeda*. Ecol Lett. 2000; **3**: 329–339.

[124] Alfaro RI. An induced defense reaction in white spruce to attack by the white-pine weevil, *Pissodes strobi*. Can J For Res. 1995; **25**: 1725–1730.

[125] Alfaro RI, Kiss GK, Yanchuk A. Variation in the induced resin response of white spruce, *Picea glauca*, to attack by Pissodes strobi. Can J For Res. 1996; **26**: 967–972.

[126] Tomlin ES, Alfaro RI, Borden JH, He FL. Histological response of resistant and susceptible white spruce to simulated white pine weevil damage. Tree Physiol. 1998; **18**: 21–28.

[127] Byun McKay SAB, Hunter WL, Godard KA, Wang SW, Martin DM, Bohlmann J, Plant AL. Insect attack and wounding induce traumatic resin duct development and gene expression of (−)-pinene synthase in *Sitka spruce*. Plant Physiol. 2003; **133**: 368–378.

[128] Martin D, Tholl D, Gershenzon J, Bohlmann J. Methyl jasmonate induces traumatic resin ducts, terpenoid resin biosynthesis, and terpenoid accumulation in developing xylem of Norway spruce stems. Plant Physiol. 2002; **129**: 1003–1018.

[129] Miller B, Madilao LL, Ralph S, Bohlmann J. Insect-induced conifer defense: white pine weevil and methyl jasmonate induce traumatic resinosis, de novo formed volatile emissions, and accumulation of terpenoid synthase and octadecanoid pathway transcripts in *Sitka spruce*. Plant Physiol. 2005; **137**: 369–382.

[130] Krekling T, Franceschi VR, Krokene P, Solheim H. Differential anatomical responses of Norway spruce stem tissues to sterile and fungus infected inoculations. Trees. 2004; **18**: 1–9.

[131] Christiansen E, Krokene P, Berryman AA, Franceschi VR, Krekling T, Lieutier F, Lönneborg A, Solheim H. Mechanical injury and fungal infection induce acquired resistance in Norway spruce. Tree Physiol. 1999; **19**: 399–403.

[132] Krokene P, Solheim H, Krekling T, Christiansen E. Inducible anatomical defense responses in Norway spruce stems and their possible role in induced resistance. Tree Physiol. 2003; **23**: 191–197.

[133] Nagy N, Franceschi VR, Solheim H, Krekling T, Christiansen E. Wound-induced traumatic resin duct development in stems of Norway spruce (Pinaceae): anatomy and cytochemical traits. Am J Bot. 2000; **87**: 302–313.

[134] Hietala AM, Kvaalen H, Schmidt A, Jøhnk N, Solheim H, Fossdal CG. Temporal and spatial profiles of chitinase expression by Norway spruce in response to bark colonization by *Heterobasidion annosum*. Appl Environ Microbiol. 2004; **70**.

[135] Staswick PE, Tiryaki I. The oxylipin signal jasmonic acid is activated by an enzyme that conjugates it to isoleucine in Arabidopsis. Plant Cell. 2004; **16**:2117–2127.

[136] Walley JW, Kliebenstein DJ, Bostock RM Dehesh K. Fatty acids and early detection of pathogens. Curr Opin Plant Biol. 2013; **16**: 520–526. DOI: 10.1016/j.pbi.2013.06.011.

[137] Zhang Q, Xiao S. Lipids in salicylic acid mediated defense in plants: focusing on the roles of phosphatidic acid and phosphatidylinositol 4-phosphate. Front Plant Sci. 2015; http://dx.doi.org/10.3389/fpls.2015.00387.

Chapter 2

Herbivore Adaptations to Plant Cyanide Defenses

Maike van Ohlen, Anna-Maria Herfurth and
Ute Wittstock

Additional information is available at the end of the chapter

http://dx.doi.org/10.5772/66277

Abstract

As plants are fixed to their habitat they produce specialized metabolites as chemical defenses to fight off herbivores. As an example, many plants produce cyanogenic glucosides and release toxic cyanide upon tissue damage ("cyanide bomb"). As a prerequisite for exploring cyanogenic plants as hosts, herbivores have evolved mechanisms to overcome cyanogenic defenses. Mammals metabolize cyanide to thiocyanate by rhodaneses. In arthropods, both rhodaneses and β-cyanoalanine synthases which transfer cyanide to cysteine contribute to cyanide detoxification. However, based on enzyme activity tests some arthropod species possess only one of these activities, and some possess both. Recently, cloning and characterization of first arthropod β-cyanoalanine synthases provided evidence for their involvement in cyanide detoxification. Phylogenetic analyses suggest that they have been recruited from microbial symbionts. Investigations with *Zygaena filipendulae* revealed that the avoidance of cyanide release is the primary mode of overcoming cyanide in this specialist. Some herbivores are able to sequester, *de novo* synthesize, and store cyanogenic glucosides for their defense and as nitrogen source. Thus, herbivores have evolved various mechanisms to counteract host plant cyanide defenses. These mechanisms are likely to have played a key role in the evolution of plant-herbivore interactions as well as in speciation and diversification of arthropods.

Keywords: cyanide detoxification, plant secondary/specialized metabolism, cyanogenic glucosides, β-cyanoalanine synthase, rhodanese

1. Introduction

Herbivores are a main threat for plants as their feeding destroys vegetative and generative parts of the plant, that is, organs needed for assimilation, nutrient storage and reproduction. In order to cut their losses, plants have developed physical and chemical defenses to fight off herbivores and to survive in their ecosystem. Very effective means to defend against predators are provided by the so-called specialized (or "secondary") metabolism, which is not required for growth and development, but for the plant's interaction with its environment [1, 2]. Specialized metabolism is the source of diverse low molecular weight compounds such as alkaloids, terpenes, glucosinolates or cyanogenic glucosides, which are often specific to certain families or species. These compounds may repel the potential predator before contact or harm the herbivore upon ingestion. Defensive metabolites may have herbivore-specific effects or be universally toxic. In the latter case, they need to be stored in an inactive or nontoxic form in the plant to avoid self-intoxication.

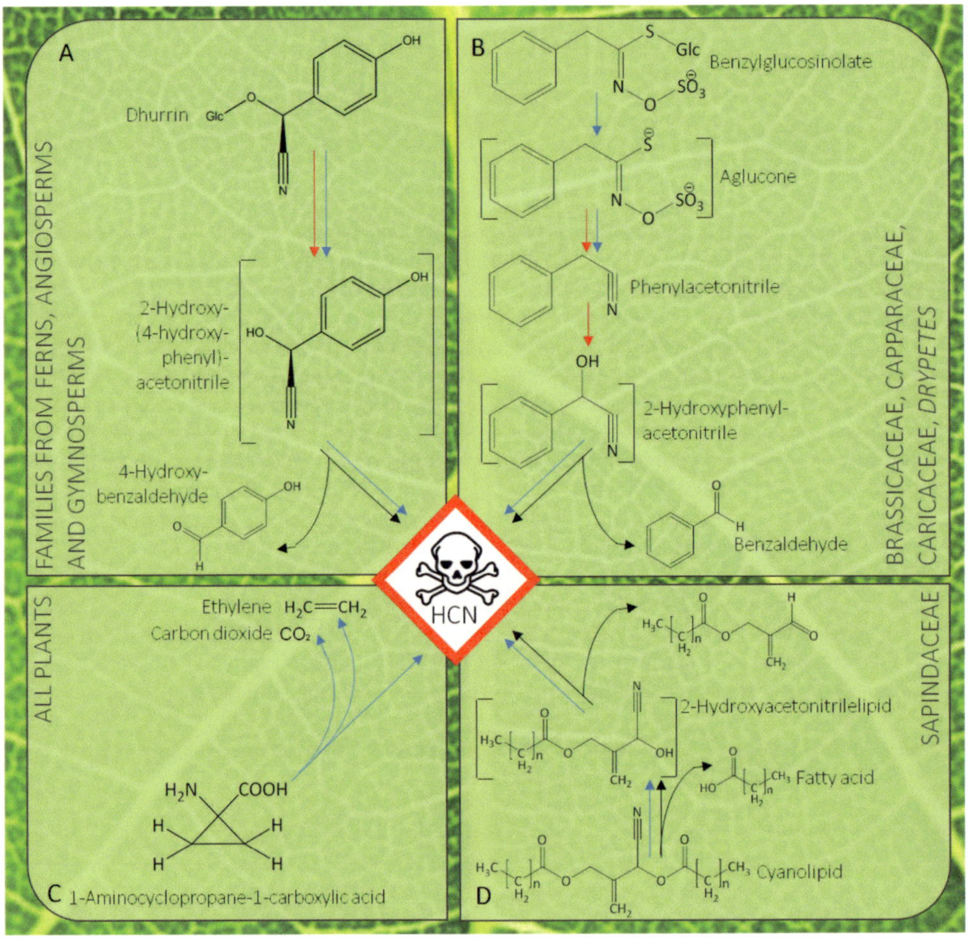

Figure 1. Sources of cyanogenesis upon herbivory. Exemplary precursors, intermediates and reactions leading to the liberation of (hydrogen)cyanide from the four main pathways are shown. (A) Cyanogenic glucoside hydrolysis, (B) metabolism of aromatic glucosinolates in the herbivore *P. rapae*, (C) final step of the ethylene biosynthesis pathway, (D) cyanolipid hydrolysis. Plant families and groups forming the precursor compounds are shown in gray. Reactions involving plants are illustrated by blue arrows, reactions involving herbivore proteins by red arrows and spontaneous reactions by black arrows. For references and details, see the main text.

Chemical defense through cyanide is widespread in the plant kingdom. As a universal respiration toxin, cyanide is not accumulated in free form in plants but released from cyanogenic precursors upon tissue damage ("cyanide bomb"), in the course of metabolic reactions in intact plant tissue or upon ingestion by herbivores (**Figure 1**). The acute and universal toxicity of cyanide in combination with its frequent occurrence in the plant kingdom calls for efficient cyanide detoxification mechanisms in herbivores. As soon as cyanide is liberated upon ingestion of cyanogenic plant material, an enzymatic detoxification is vital for the protection of the herbivore's cellular metabolism. Although diverse enzymatically catalyzed reactions for the detoxification of cyanide have been described in microorganisms [3], only two main pathways of cyanide detoxification are present in higher animals. These are on the one hand the rhodanese or 3-mercaptopyruvate sulfurtransferase-catalyzed transfer of sulfur from a donor substrate to cyanide, leading to the formation of thiocyanate, and, on the other hand, the β-cyanoalanine synthase-catalyzed substitution of cysteine's sulfhydryl group by cyanide, leading to the formation of β-cyanoalanine.

An efficient way to minimize the risk of cyanide poisoning is to prevent its formation. Therefore, herbivorous arthropods that colonize plants with high cyanide potential often possess specialized adaptations, which allow them to avoid cyanide release upon ingestion of plant material. This chapter introduces cyanide as a ubiquitous plant-produced compound and summarizes the present understanding of cyanide detoxification pathways and the involved enzymes as well as the current knowledge on cyanide avoidance mechanisms in herbivores with a special focus on arthropods. As certain arthropod species are able to synthesize cyanogenic compounds themselves and/or to sequester cyanogenic compounds from their food plants, we also discuss cyanogenesis in herbivores from an ecological and evolutionary perspective.

2. Sources of cyanide exposure

The most common storage form of cyanide in plants is cyanogenic glucosides, which are potent antiherbivore defenses with an additional function as nitrogen storage compounds [4–8]. The intact glucosides are water-soluble and nontoxic compounds, but hydrolysis catalyzed by β-glucosidases liberates the cyanohydrins (α-hydroxynitriles), which, spontaneously or under catalysis by α-hydroxynitrilases, release the hydrogen cyanide next to aldehydes or ketones (**Figure 1A**) [9]. Their broad distribution among more than 2650 plant species from the pteridophytes, gymnosperms and angiosperms [9, 10] may be explained by their biosynthetic origin. Cyanogenic glucosides are biosynthesized through oxidation of common aliphatic and aromatic amino acids by members of the wide-ranging cytochrome P450 family with oximes and cyanohydrins as intermediates and subsequent *O*-glycosylation [11]. Although cyanogenic glucosides are widespread within the plant kingdom and their biosynthetic enzymes are ancient [12], <30 different structures of plant-derived cyanogenic glucosides have been described. Among those, linamarin, lotaustralin and dhurrin (**Figure 1A**), derived from the amino acids valine, isoleucine and tyrosine, respectively, are the most common glucosides. Activation of the glucosides under cyanide liberation happens upon herbivore attack when the glucosidic bond is hydrolyzed by endogenous plant β-glucosidases or digestive enzymes of the herbivore (**Figure 1A**). Plants possess β-glucosidases with high specificity toward their own defensive cyanogenic glucosides [13].

Some cyanogenic glucosides have not only been found in plants, but are sequestered and even biosynthesized *de novo* in herbivores and detritivores from the Arthropoda [14, 15]. As a prominent example, larvae of *Zygaena filipendulae* (Lepidoptera: Zygaenidae) which are specialist herbivores on the cyanogenic bird's-foot trefoil *Lotus corniculatus* are able to sequester cyanogenic glucosides from the host plant and to synthesize them *de novo* by a pathway that has evolved independently from that in plants (see below) [16, 17]. Both the plant and the insect benefit from inherent cyanogenic glucosides as a chemical defense and as nutrient storage compounds [4, 6]. Thus, the Zygaena-Lotus association does not only illustrate the coevolutionary arms race between plants and their herbivores but also provides an example of convergent pathway evolution [16, 18, 19]. Cyanide intoxication from the diet is also relevant to humans as some major foodstuffs, especially in tropical regions, contain cyanogenic glucosides. In particular, cassava tubers, green sorghum leaves, white clover foliage and lima beans have been reported as potential sources of cyanide to man. Nevertheless, traditional food preparation and selective breeding to decrease cyanide levels lower the risk of accidental cyanide poisoning.

A further class of plant secondary compounds carrying a cyanide group is the cyanolipids, a group of lipids possessing a branched five carbon skeleton with a nitrile group [20]. Cyanolipids occur in the seed oil of diverse species of the Sapindaceae [21]. They are cyanohydrin esters, that is, they possess an esterified hydroxyl group in α-position to the nitrile moiety and will form unstable α-hydroxynitriles upon spontaneous or lipase-catalyzed ester hydrolysis (**Figure 1D**) [22]. As described above, α-hydroxynitriles are a source of cyanide as they readily decompose either spontaneously or enzymatically catalyzed.

Cyanide may also be liberated upon metabolism of another group of specialized metabolites, the glucosinolates, inside the herbivore by the consecutive action of plant- and herbivore-expressed enzymes. Glucosinolates are amino acid-derived thioglucosides with a sulfated aldoxime core and a variable side chain [23] (**Figure 1B**). They are part of the glucosinolate-myrosinase system or "mustard oil bomb," a constitutive defense mechanism common to all families of the Brassicales. The products arising from glucosinolate hydrolysis have manifold effects on herbivores feeding on Brassicales plants, including general deterrence and toxicity, but may also be perceived by specialist herbivores and their parasitoids as host identification cues [24]. The primary defense compounds derived from this system are the isothiocyanates which result from rearrangement of the aglucone formed upon hydrolysis by co-occurring thioglucosidases (myrosinases) when tissue is disrupted [23, 25]. Besides isothiocyanates, other products such as nitriles, epithionitriles and organic thiocyanate can also be formed depending on the structure of the glucosinolate side chain and the presence of additional plant-expressed proteins, the so-called specifier proteins [26–29]. Cyanide release from a glucosinolate-derived nitrile has been demonstrated to occur in larvae of *Pieris rapae* (Lepidoptera: Pieridae) which are specialized feeders on glucosinolate-containing plants [30] (**Figure 1B**). Larvae of *P. rapae* and other glucosinolate-feeding Pieridae produce a gut nitrile-specifier protein to overcome the glucosinolate defense of their host plants [31, 32]. Upon ingestion of glucosinolate-containing plant material, this protein redirects glucosinolate hydrolysis catalyzed by plant myrosinases entirely toward nitriles (instead of isothiocyanates) which are excreted or further metabolized [33, 34] (**Figure 1B**). Phenylacetonitrile (derived from benzylglucosinolate) undergoes α-hydroxylation by microsomal *P. rapae* enzymes yielding a cyanohydrin which is subsequently decomposed to cyanide and an aldehyde [30] (**Figure 1B**). Despite liberation of one cyanide molecule per molecule of ingested benzylglucosinolate, the larvae feed on benzylglucosinolate-containing plants without ill effects indicating adaptation to this toxin [30]. Expression of a gut nitrile-specifier protein appears to be confined to Pieridae species, while glucosinolate-feeding species of other families overcome the glucosinolate-myrosinase system by other means [32, 35]. If herbivores devoid of a gut nitrile-specifier protein may also encounter cyanide liberated from glucosinolate breakdown products is presently uncertain. However, cyanide release from glucosinolate metabolites in homogenates of *Alliaria petiolata* (Brassicaceae) [36], nitrile formation upon glucosinolate breakdown in plant homogenates [27, 28, 37] and the likely ability of herbivore phase-I-detoxification enzymes to hydroxylate these nitriles to cyanohydrins [38] indicates that herbivores outside the Pieridae are likely exposed occasionally to cyanide when feeding on glucosinolate-producing plants.

Apart from the accumulation of cyanogenic precursors as part of specialized metabolism for defense against herbivores, plants from all families generate cyanide as a by-product during the formation of ethylene, a ubiquitous plant hormone. In the last step of ethylene biosynthesis, the oxidation of 1-aminocyclopropane-1-carboxylic acid to ethylene, one mole of cyanide is liberated per mole of ethylene formed (**Figure 1C**) [39]. Although the steady-state concentration of cyanide from this pathway is normally kept at a low level of 0.2 μM by action of cyanide detoxifying enzymes [40], this demonstrates the ubiquitous occurrence of cyanide in the feed of herbivores.

Taken together, cyanide is universally present in plants and herbivores are frequently confronted with this toxin through their diet. Thus, safe handling of cyanide is a necessary prerequisite for herbivory by both specialists feeding exclusively on cyanide-defended plants and generalists with occasional cyanide ingestion. Nevertheless, a varying cyanide content in the host plant seems to influence generalist herbivory more severely than specialist feeding indicating the existence of efficient adaptations to deal with this toxin [41].

3. Cyanide toxicity

Uptake of the small and simple ion cyanide has tremendous effects on the metabolism of all aerobic cells, resulting from its high reactivity and efficient binding to various proteins of cellular respiration and regulation. The main reason for its acute and universal toxicity is the formation of stable complexes between cyanide and the Fe^{3+}-ion of heme a_3 of cytochrome c oxidase, one of the electron carriers in the respiratory chain. Cyanide binds to cytochrome c oxidase and acts as a noncompetitive inhibitor of cytochrome c. This stops electron transfer, leading to termination of the respiratory chain and the citric acid cycle due to a shortage of the electron acceptor NAD^+ [42]. The resulting lack of ATP is detrimental to the cell. As a consequence, glycolysis, the alternative, but inefficient pathway of ATP generation, is accelerated in combination with lactic acid fermentation for regeneration of NAD^+. In humans, metabolic acidosis resulting from high lactic acid levels is responsible for most of the symptoms of cyanide intoxication [43]. Besides the Fe^{3+} of cytochrome c oxidase cyanide binds metal ions of various metalloenzymes, in particular molybdoenzymes, and forms Schiff base intermediates with pyridoxal phosphate-dependent enzymes causing an efficient inhibition of a wide range of metabolic reactions and regulatory processes in the cell [44].

In vertebrates, cyanide does not only influence cellular metabolism but also diverse physiological processes. By binding to chemoreceptors, cyanide causes vasoconstriction of main arteries which may lead to cardiac shock or pulmonary edema [43, 45]. In addition, cyanide may increase neurotransmitter release by influencing calcium channels in neural cell membranes [46]. Even sublethal doses of cyanide may harm the brain of mammals by altering the membrane lipid peroxidation and the response of antioxidant enzymes [47].

Several studies performed by Edwin J. Bond in the 1960s on the beetle *Sitophilus granarius* have investigated the effect of cyanide on insects as discussed by Page and Lubatti [48]. The studies showed that consumption of cyanide even at low doses results in immediate paralysis of the animals by reduced respiration [48, 49]. Lethal effects are not only due to cytochrome c oxidase inhibition but also result from increased proteolysis and binding of cyanide to intermediates of glycolysis. In contrast to mammals, where supplementation of oxygen is an efficient treatment of cyanide poisoning, the administration of oxygen to cyanide-exposed insects amplifies the ill effects of cyanide, probably due to the accumulation of promptly formed peroxides and acidosis-causing citrate and pyruvate. In contrast, oxygen exclusion allows the animals to recover from the poisoning [50, 51].

Thus, the mode of action of the poison cyanide is complex, and the lethal effects differ between species. Nevertheless, cyanide is one of the most potent toxins and an efficient and universal weapon of plants against herbivore foraging.

4. Cyanide detoxification enzymes

4.1. Sulfur transferases

Sulfur transferases such as the rhodanese (thiosulfate:cyanidesulfurtransferase, EC 2.8.1.1; see **Figure 2A**), and its close relative, the 3-mercaptopyruvate sulfurtransferase (EC 2.8.1.2), are enzymes described in plants, fungi, bacteria and a wide range of animals including snails, insects, fish and mammals (see **Figure 3**) [52–58]. Enzymatic formation of thiocyanate, the so-called rhodanide, was first described in 1933 using vertebrate tissues as discussed by Lewis [59]. Rhodaneses from mammals have been investigated most thoroughly and most insight has been gained from the examination of human and bovine liver rhodaneses [60, 61]. These two enzymes served to uncover the first protein structure of a rhodanese which revealed two similarly folded "rhodanese domains" [62]. In contrast to the highly similar tertiary structure, the two domains differ strongly in their amino acid sequences in agreement with their diver-

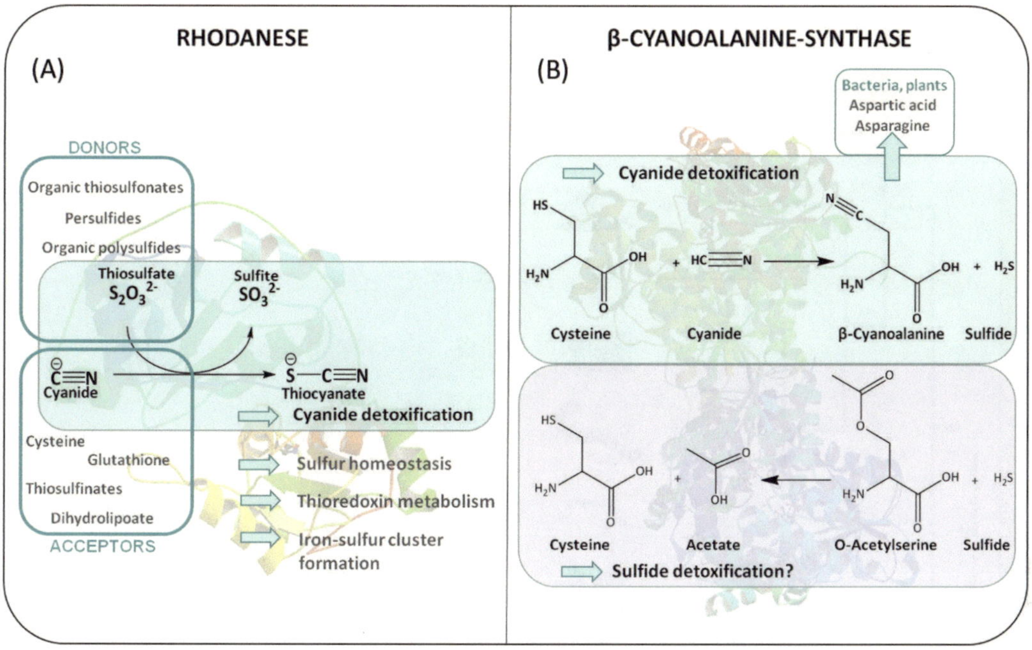

Figure 2. Main cyanide detoxification enzymes. Shown are reactions in which cyanide conversion is catalyzed by rhodanese (A—background: bovine liver rhodanese PDB 1RHD) or β-cyanoalanine synthase (B—background: soy bean β-cyanoalanine synthase PDB 3VBE). Additional rhodanese substrates are also shown (A). The *O*-acetylserine thiol lyase reaction is shown in the lower panel of B. Block arrows indicate possible alternative and additional roles of the enzymes.

gent functions as C-terminal catalytical and N-terminal regulatory domains [62]. At the level of primary structure, rhodaneses from different organisms show little similarity apart from two conserved, 11–13 amino acids long "signature" regions at the N- and C-termini which are also present in distantly related proteins of the rhodanese superfamily such as cdc25 phosphatase and heat shock proteins [63]. This low identity at the amino acid level and the involvement of single rhodanese domains in aberrant proteins make homology-based identification of rhodaneses from further species difficult. To specify the sulfur transferases involved in cyanide detoxification among the members of the diverse superfamily, the tertiary structure has to be taken into account, classifying the true rhodaneses as tandem domain thiosulfate:cyanide sulfurtransferases [63]. This group also comprises the 3-mercaptopyruvate sulfurtransferases. Both types of enzymes have distinct substrate and product spectra, but are yet interconvertible by few amino acid substitutions [64].

Rhodaneses do not only accept thiosulfate as sulfur donor, but all sulfane anions such as organic sulfanes and persulfides [55, 65]. Next to cyanide, the sulfur atom may be accepted by other thiophilic substrates such as the amino acids cysteine and glutathione [65]. The kinetic mechanism of rhodanese was uncovered with its classical substrates cyanide and thiosulfate by Westley and coworkers [60, 66]. In a ping-pong reaction, the sulfane sulfur atom is abstracted from the donor substrate thiosulfate and bound to a cysteine residue in the active site of rhodanese [62]. This is followed by entrance of the acceptor substrate cyanide into

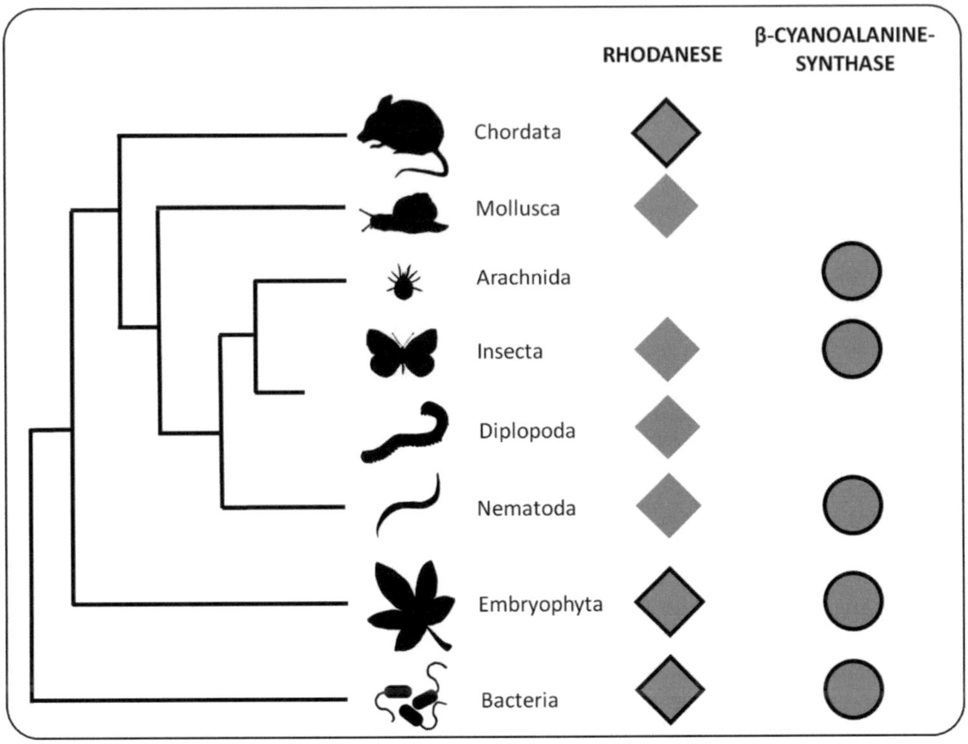

Figure 3. Occurrence of cyanide detoxification enzymes in living organisms. Selected domains, subkingdoms, phyla and classes are shown in a schematic representation of their phylogenetic relationship. Groups in which enzyme activity has been detected are labeled with a gray area (square for rhodanese, circle for β-cyanoalanine synthase). The area is surrounded by black line if sequences of the corresponding enzymes or their genes have been elucidated. Metabolite data also proved β-cyanoalanine synthase activity in Diplopoda, where activity assays have not been performed to our knowledge. Rhodanese seems to be an ubiquitous enzyme, although no sequence data are available from Arthropoda or Mollusca. In contrast, β-cyanoalanine synthase was detected primarily in plants, bacteria and Arthropoda.

the active site and transfer of the sulfur atom [60, 66]. In contrast, the reaction of 3-mercaptopyruvate sulfurtransferase (which can also convert cyanide to thiocyanate [67]) follows a sequential mechanism with formation of a ternary complex composed of the enzyme and both substrates (3-mercaptopyruvate and cyanide) [68, 69].

A main function of the rhodaneses in cyanide detoxification is in agreement with their subcellular localization. Rhodanese activity is predominantly detected in the mitochondria, the site of cellular respiration with the cyanide-susceptible cytochrome c oxidase [70, 71]. Nevertheless, for 3-mercaptopyruvate sulfurtransferase and rhodanese of some species, an additional localization in the cytosolic fraction has been described [55, 71, 72]. The cytosolic enzymes may serve to reduce cyanide levels in this compartment whose components (glycolysis intermediates, proteins) may also be affected by cyanide poisoning (see above) [48]. The reaction product of rhodanese-catalyzed cyanide detoxification, thiocyanate, possesses a toxic potential toward mitochondria which could be an additional reason why rhodanese isoforms are also localized in the cytosol in several species.

Although a major role of rhodaneses in cyanide detoxification seems likely based on the present knowledge, cyanide detoxification might not be their exclusive physiological function. The ubiquitous occurrence of rhodaneses in organisms and tissues with no obvious cyanide exposure as well as the low physiological concentration of their substrate thiosulfate in the mitochondria has fueled doubts about their main role in cyanide detoxification [73]. In support of a major role in cyanide detoxification in mammals, rhodanese activity is inducible in rats by exposition to cyanide or supplementation with thiosulfate [74, 75]. *In vivo*, sulfane substrates may be provided by the action of an enzyme involved in cysteine metabolism in animals, cystathionine-γ-lyase [76]. Further and alternative functions of rhodaneses have been discussed. As plant homologs have been shown *in vitro* to supply sulfide for iron sulfur clusters of the electron transport chain, this has been suggested as a general function of rhodaneses from other taxa [73, 77]. Alternatively, a general role of rhodaneses in the regulation of sulfur homeostasis has been proposed based on their low substrate specificity. The *in vivo* function of 3-mercaptopyruvate sulfurtransferases in cyanide detoxification has been questioned not least because addition of 3-mercaptopyruvate led to an antidote effect only in some mammal species [78]. Additionally, it has been difficult to separate their possible direct and indirect contributions to cyanide detoxification *in vivo*, as 3-mercaptopyruvate sulfurtransferase can also donate sulfur for the formation of the rhodanese substrate thiosulfate [79].

4.2. β-Cyanoalanine synthase

The main enzyme of cyanide detoxification in plants and many bacteria, β-cyanoalanine synthase (EC 4.4.1.9, **Figures 2B** and **3**), belongs to the family of β-substituted alanine synthases sharing the cofactor pyridoxal-5′-phosphate and a uniform fold [80, 81]. This family also comprises *O*-acetylserine thiol lyase and cystathionine-β-synthase which occur in bacteria, plants, fungi and animals and are involved in cysteine biosynthesis from *O*-acetylserine or serine and homocysteine, respectively [80, 82]. The amino acid sequence of these enzymes and their tertiary structure, the so-called fold II of the pyridoxal-5′-phosphate-dependent enzymes, are highly conserved. However, the members of the protein family vary in the quaternary structure, as mono-, di-, tetra- and oligomers have been reported, although dimers are most predominant [80, 83, 84]. Some members of the family can catalyze several of the above-mentioned reactions making a classification as β-cyanoalanine synthase or *O*-acetylserine thiol lyase difficult. Thus, plant enzymes are mainly assigned to an enzyme clade based on a comparison of kinetic characteristics for the diverse reactions catalyzed and their subcellular localization [80]. The only phyla of multicellular animals in which β-cyanoalanine synthase activity has been demonstrated up to now are Nematoda and Arthropoda (see **Figure 3**) [14, 85–87]. Only few animal β-substituted alanine synthases with β-cyanoalanine synthase activity have been identified to date (see below). As the protein family's name indicates, β-substituted alanine synthases catalyze a substitution or elimination reaction at the β-carbon of proteinogenic and nonproteinogenic amino acids. Their reaction mechanism has been examined mainly on the basis of prokaryotic and plant *O*-acetylserine thiol lyases [88–90]. All examined β-substituted alanine synthases share a common fold and perform a similar conformational change upon substrate binding to close the active site for catalysis [90]. In the active site, a Schiff basebinding between the cofactor pyridoxal-5′-

phosphate and a catalytic lysine residue awaits the substrate amino acid [90]. Upon entrance of the substrate, pyridoxal-5′-phosphate is transferred from the lysine residue to the α-amino group of the substrate as an external Schiff base. Lysine's free amino group now acts as a base catalyst for the deprotonation of the α-carbon, inducing the α,β-elimination of sulfide, acetate or water from the substrate amino acid [88, 90]. The formed α-aminoacrylate intermediate is shared by all members of the β-substituted alanine synthase family regardless of the catalyzed reaction [88]. After formation of this intermediate, the second substrate enters the active site. It attacks the amino acid's side chain nucleophilically. Facilitated by acid catalysis by the protonated lysine residue, a bond between the β-carbon and the second substrate is formed. The newly formed amino acid product can then be released from the active site under reformation of the internal Schiff base between pyridoxal-5′-phosphate and lysine [90]. Because of high conservation in the reaction mechanism of different β-substituted alanine synthases, an *O*-acetylserine thiol lyase and a β-cyanoalanine synthase from soybean were analyzed for their substrate and product spectra by mutagenesis studies [90]. This showed that the exchange of only three amino acids can switch an *O*-acetylserine thiol lyase to a β-cyanoalanine synthase, illustrating their close relationship and giving insight into the evolution of β-substituted alanine synthases [90].

Next to amino acid biosynthesis, β-substituted alanine synthases are involved in cellular sulfur and redox homeostasis [84, 91–93]. Cyanide detoxification by these enzymes is mainly catalyzed by β-cyanoalanine synthases, but *O*-acetylserine thiol lyases can also bind the toxic ion to either cysteine or *O*-acetylserine [81]. According to enzyme activity assays, β-cyanoalanine synthases from plants and animals are mainly localized in the mitochondria [86, 94]. However, the recently identified arthropod β-cyanoalanine synthases do not seem to possess a mitochondrial signal sequence [87, 95]. Cytosolic expression might be beneficial as cyanide detoxification by β-cyanoalanine synthases leads to the formation of equimolar amounts of sulfide which itself is an inhibitor of cytochrome c oxidase. If released in the cytosol (instead of the mitochondria), sulfide can immediately be captured by the cytosolic *O*-acetylserine thiol lyases. In plants, experiments with T-DNA insertion mutants have shown that cytosolic *O*-acetylserine thiol lyases are essential for safe and efficient disposal of cyanide [92].

β-Cyanoalanine itself may also exert harmful effects. It has been identified as a neurotoxin and can also be lethal to plants [96, 97]. In order to protect themselves from poisoning with β-cyanoalanine and to minimize costs, plants and microorganisms are able to turn over β-cyanoalanine by nitrile hydratases and nitrilases [98]. Nitrile hydratases catalyze the addition of a water molecule to β-cyanoalanine leading to the formation of the proteinogenic amino acid asparagine. Nitrilases convert β-cyanoalanine to the proteinogenic amino acid aspartate by addition of two water molecules [99]. In addition, the conversion of β-cyanoalanine to asparagine with γ-glutamyl-β-cyanoalanine as an intermediate has been described for some plants [96]. Recycling of β-cyanoalanine has also been shown in some arthropod species (see below). Due to its neurotoxic effect, β-cyanoalanine stored in the defensive droplets of the cyanogenic lepidopteran species *Z. filipendulae* has also been discussed to directly act as a defensive compound for the protection of this herbivore against predators [100].

5. Cyanide detoxification strategies in herbivores

5.1. Cyanide detoxification in mammalian herbi- and omnivores

In mammals, rhodanese is generally believed to be the major enzyme for cyanide detoxification, while β-cyanoalanine synthase activity has not been detected in a mammal so far. A comparison of rhodanese activity between mammalian herbi-, omni- and carnivores shows highest activities in herbivores, especially in ruminants which feed on a broad range of plant material including plants with high cyanide potential [58]. In several mammalian species such as plant-feeding rabbits, rhodanese activity is ubiquitously distributed in the body with the highest activity in hepatocytes, the main detoxification site of e.g. xenobiotics [101, 102]. Rhodanese activity is also localized in the mammalian brain, where cyanide acts as a neuromodulator [63].

5.2. Cyanide detoxification in invertebrates

Intensive research on mammalian rhodaneses also raised the question whether these enzymes are involved in cyanide detoxification in other animals. Rhodanese activity is widely distributed in insects and occurs in snails (see **Figure 3**) [55, 103]. The level of activity is comparable to that of mammalian gut tissue [53]. However, activity levels are largely in the same range among herbivores which frequently or rarely encounter high cyanide levels. The basal rhodanese activity might be sufficient to capture dietary cyanide in herbivores regardless of the cyanide level in the diet. Alternatively, the uniform distribution of rhodanese activity among herbivores could indicate that arthropod rhodaneses possess an additional function unrelated to cyanide detoxification. This would likely require other mechanisms of cyanide detoxification such as β-cyanoalanine synthase activity [103]. In agreement with this, β-cyanoalanine activity has been found to be broadly distributed in arthropod herbivores (see **Figure 3**) [86, 95, 104]. In support of the activity data, labeled β-cyanoalanine can be detected in arthropods after feeding of or exposition to isotopically labeled cyanide [30, 105]. Further support for the relevance of β-cyanoalanine synthases comes from experiments with several cyanide-forming lepidopteran species in whose defensive glands β-cyanoalanine and its hydration product asparagine were detected [106]. In millipedes, β-cyanoalanine synthase-catalyzed detoxification of cyanide and further metabolism of β-cyanoalanine to asparagine was demonstrated by studies with radiolabeled precursors [14, 107]. For insects, a similar utilization of cyanide for the formation of proteinogenic amino acids using β-cyanoalanine as an intermediate has been discussed [18], but has only be proven for one beetle by radioactive feeding experiments so far [108]. In the beetle, the radioactive label was recovered from a polypeptide rich in aspartate, the product of nitrilase-catalyzed conversion of β-cyanoalanine [108].

In order to estimate the relevance of rhodanese *vs.* β-cyanoalanine synthase for the *in vivo* detoxification, several studies compared the enzyme activities within one species upon cyanide induction or in related species with or without cyanide specialization [53, 86]. Most of these studies are more than 20 years old, and they cover only a very limited number of species. Nevertheless, for rhodanese no induction of activity upon cyanide feeding was observed although the occurrence in Diptera species parasitizing the cyanogenic moth *Z. filipendulae* indicates a connection with cyanide detoxification [109]. β-Cyanoalanine synthase activity was higher in cyanide-tolerant than in cyanide-sensitive species [86]. Three studies analyzed a broad spectrum of insect species either by measuring rhodanese enzyme activities or by determining β-cyanoalanine as an indicator of β-cyanoalanine synthase activity [103, 106, 110]. However, in addition to the difficulty to compare enzyme activity and metabolite level *per se*, several other factors render a statement on the *in vivo* relevance of β-cyanoalanine synthase and rhodanese for cyanide detoxification in insects impossible: the low number of replicates [103], the use of specimen of different developmental stages and from different origins [103, 110], quantification by thin-layer chromatography, and long-term storage of specimen before analysis despite decomposition of metabolites over time [106, 110]. Thus, conclusions on the quantitative contribution of the two pathways of cyanide detoxification in insects cannot be drawn from these experiments. Nevertheless, the studies showed that both enzymes have a broad distribution among insects and many species possess both activities [30, 55, 103, 104, 110, 111]. β-Cyanoalanine synthase seems to be most relevant to protect arthropod herbivores feeding on cyanide-defended plants.

Most β-cyanoalanine synthase activity data for animals were generated with intact, partly dried and stored animal tissues, but some enzymes have been purified and characterized [84, 104]. In the nematode *Caenorhabditis elegans*, identification of the first animal members of the β-substituted alanine synthase family that do not belong to the cystathionine-β-synthases was achieved at the molecular level [85]. Biochemical characterization of the enzymes proposed roles in the regulation of the metabolism and the detoxification of cyanide, sulfide and S-sulfocysteine [84]. Although the enzymes possess *O*-acetylserine thiol lyase activity *in vitro*, an *in vivo* function in sulfur assimilation and cysteine biosynthesis is unlikely [84]. Further β-cyanoalanine synthases have been identified in the mite *Tetranychus urticae* and larvae of *P. rapae* [87, 95]. These first molecular data indicate an astonishing evolutionary background. In phylogenetic analyses, the nematode enzymes group together with plant *O*-acetylserine thiol lyases [84, 85] indicating a common origin of β-cyanoalanine synthases of both kingdoms, that is, acquisition of β-cyanoalanine synthases by gene duplication and partial neofunctionalization of a common ancestor. In contrast, the β-cyanoalanine synthases from the mite and the butterfly show amino acid sequence similarity with bacterial sequences [87, 95]. Phylogenetic analyses group the mite and butterfly enzymes among bacterial homologs from α- and β-proteobacteria, distant to other metazoan β-substituted alanine synthases [87, 95]. Genetic analyses such as determination of the GC-nucleotide content and mapping of adjacent genomic DNA have shown that these sequences belong to the arthropods' genome rather than to the genome of bacterial symbionts of the present species [87, 95]. As a likely explanation, the genes may have been acquired by horizontal gene transfer from bacteria, likely from symbiotic bacteria which lived in close association with ancestral arthropods [87]. Nevertheless, a common horizontal gene transfer event to arthropods seems to be unlikely as this would involve at least 13 independent gene losses each at a very specific point in the evolution [87]. Two independent gene transfer events from closely related bacterial donor species or a gene transfer from bacterium to mite followed by a second transfer event from an ancient mite to a Lepidopteran ancestor are discussed [87]. Interestingly, mites possess only one copy of the sequence in their genome, while some Lepidoptera seem to possess two or three copies [95], a phenomenon frequently observed for genes assimilated by horizontal gene transfer [112]. Horizontal gene transfer, in particular from bacteria and protists, but also from plants and fungi, has contributed gravely to metazoan evolution. Its remnants besides the β-cyanoalanine synthases can be found in tens to hundreds of examples in nematodes, arthropods and chordates and are involved in main metabolic pathways and responses to environmental influences [113].

In general, proteins involved in the adaptation of herbivores to their host plants and, in particular, those catalyzing the detoxification and transport of host plant xenobiotics are thought to be under narrow transcriptional regulation [114]. For the mite β-cyanoalanine synthase, an induction by cyanide exposure over 30 generations led to a transcriptional response allowing for the identification of the detoxification enzyme [87]. Thus, these enzymes are among the most variable ones and play a key role in the adaptation to and population of new host plants [114]. As the relationship between herbivore and host plant is close and evolution favors adapted defense of the plant in order to diminish resource and tissue loss through predation [115], transcriptional responses could be discovered also in the host plant [114]. This close coevolution between herbivores and their host plants has therefore shaped both partners and is likely to underlie the higher cyanide tolerance of specialist herbivores on cyanogenic plants [30, 115].

6. Alternative herbivore strategies to cope with cyanogens

Next to efficient means of cyanide detoxification, herbivores have developed alternative ways to avoid intoxications when feeding on cyanide-defended plants. Often, the cyanide potential of food plants is below a toxic threshold [116]. As generalist herbivores usually change their food plants frequently, this allows them to mix a cyanide-rich diet with a diet low in cyanide to keep the overall cyanide intake below a toxic threshold [116]. Moreover, cyanogenic glucoside occurrence is often accompanied by a bitter taste of the potential food plant and many herbivores therefore avoid feeding on these plants if other host plants are available [116]. Nevertheless, in no-choice feeding experiments or if no other food plant is available in the habitat, herbivores may consume high amounts of cyanide-defended plants leading to intoxication or even death [116, 117]. Adaptations to reduce this risk include morphological, behavioral, physiological and biochemical mechanisms as outlined in the following paragraphs.

The mouthparts of herbivores from the Aphididae have evolved to specialized sucking styli which they insert through the apoplast into the sieve elements to suck phloem sap. This feeding mode avoids tissue disruption and therefore the mixing of plant cyanogenic glucosides and their spatially separated hydrolysis enzymes [41, 116, 118, 119]. In the lepidopteran specialist *Z. filipendulae* feeding on the cyanogenic glucoside-rich *L. corniculatus*, a leaf-snipping mode (leaving large portions of the plant tissue intact) in combination with a high feeding speed (shortening the time of a potential interaction between plant β-glucosidases and their substrates) decreases cyanogenic glucoside hydrolysis [120]. A similar leaf-snipping mode to avoid large-scale tissue damage was observed in a range of lepidopteran larvae not specialized on plants containing cyanogenic glucosides, including generalists as well as specialists on plants possessing another activated defense system, the glucosinolate-myrosinase-system [118]. All tested species tolerated high cyanogenic glucoside levels in their diet and excreted intact cyanogenic glucosides with their frass [118]. This was mainly achieved by avoiding cyanogenic glucoside hydrolysis through the feeding mode [118] indicating that leaf-snipping might be an ancient trait which evolved and has been maintained in lepidopteran herbivores in response to the frequent occurrence of activated plant defenses.

As a physiological adaptation, the strongly alkaline midgut pH found in some generalist and specialist herbivores allows for the inhibition of the ingested plant β-glucosidases and avoidance of cyanide liberation [118, 120] in contrast to other species with a slightly acidic midgut pH that are prone to cyanide intoxication [121, 122]. Further, properties and expression of the herbivore's endogenous β-glucosidases have undergone adaptational adjustments to reduce cyanide release from ingested plant material. As an example, the β-glucosidases localized in the saliva and midgut lumen of the cyanogenic glucoside-feeding specialist *Z. filipendulae* have lost their activity toward their host plants' cyanogenic glucosides, linamarin and lotaustralin [120]. Larvae of *Diatraea saccharalis* (Lepidoptera:Crambidae) are adapted to a cyanogenic diet by reducing the expression of aryl β-glucosidase in their midgut, thus decreasing cyanogenic glucoside catabolism and cyanide liberation [123]. A unique alternative strategy to avoid cyanide release which relies on metabolism of the cyanogenic glucoside before hydrolysis has only been described for larvae of *Heliconius sara* (Lepidoptera:Nymphalidae) so far [124]. Larvae of *H. sara* feed exclusively on leaves of *Passiflora auriculata* (Passifloraceae) with cyanocyclopentenyl glucosides as major cyanogens. Upon ingestion of leaf material, the nitrile group of the main cyanogenic glucoside is specifically replaced by a thiol to produce a compound which is not a cyanide precursor anymore [124].

Yet another mechanism protects millipede species (Diplopoda) from cyanide poisoning. These animals possess a highly tolerant cytochrome c oxidase, making cyanide poisoning less effective [125]. Instead of or in combination with a cyanide-resistant terminal oxidase, a complete cyanide-insensitive oxidative pathway has also been proposed [126]. Studies on the respiratory rate of larvae of the lepidopteran generalist herbivore, *Spodoptera eridania* (Lepidoptera:Noctuidae), also showed high cyanide tolerance, possibly indicating insensitivity of their terminal oxidase [126]. Other mechanisms of cyanide tolerance may await their discovery. In addition, a possible contribution of enzyme activities from closely associated anaerobic cyanide-resistant gut microbia has to be clarified [127].

7. Cyanogenic compounds in arthropod herbivores

7.1. Occurrence of cyanogenic compounds in arthropods

Arthropods are the only phylum of animals in which biosynthesis or sequestration of cyanogenic compounds has been shown [15]. Within arthropods, the presence of cyanogenic glucosides appears to be restricted to millipedes (Diplopoda), centipedes (Chilopoda) and three orders within the Insecta (Lepidoptera, Coleoptera and Hemiptera) [128]. The Lepidoptera and Hemiptera are the only groups containing cyanogenic compounds with aliphatic side chains, while in the others groups of arthropods, cyanogenic compounds possess aromatic side chains [4, 129]. Among the most intensely studied species, larvae of *Z. filipendulae* specialized on cyanogenic plants are able to *de novo* biosynthesize cyanogenic glucosides, but can also sequester cyanogenic glucosides from their food plants [4, 17, 130] (see Sections 7.2 and 7.3). Both *de novo* biosynthesis and sequestration of cyanogenic glucosides have also been reported for larvae of *Euptoieta hegesia* (Lepidoptera:Nymphalidae) based on the detection of cyanide after larval feeding on acyanogenic plants and an increase of cyanide formation upon transfer to cyanogenic glucoside-defended plants [131]. Species of the genus Heliconius are also able to synthesize and to sequester cyanogenic glucosides based on the pattern of compounds detected in the insects relative to those found in the food plants [124, 132]. In contrast to lepidopterans, millipedes do not store glycosides, but unglycosylated cyanogens, and do not sequester cyanide precursors from their diet as discussed in Ref. [133].

7.2. *De novo* biosynthesis of cyanogenic compounds in arthropods

First indications for *de novo* biosynthesis of cyanogenic glucosides in a herbivore came from experiments conducted by Jones and coworkers in 1962. They showed that larvae of *Z. filipendulae* release cyanide upon tissue disruption also when raised on food plants devoid of cyanogenic glucosides [109]. Feeding experiments with ^{13}C-labeled valine and isoleucine showed the incorporation of the isotope label into linamarin and lotaustralin and thereby provided evidence for *de novo* biosynthesis of cyanogenic glucosides in *Z. filipendulae* and the butterfly *Heliconius melpomene* (Lepidoptera:Nymphalidae) [132, 134]. The complete biosynthetic pathway of cyanogenic glucosides in *Z. filipendulae* was elucidated in 2011 including pathway intermediates and involved enzymes [16, 17, 19, 134, 135]. This revealed high similarity to the biosynthesis of cyanogenic glucosides in plants with two cytochrome P450 enzymes catalyzing the conversion of the precursor amino acid to a cyanohydrin via an aldoxime and a glucosyltransferase responsible for the final glycosylation step (see **Figure 4**). Several butterfly species of the Papilionoidae also contain linamarin and, in part, lotaustralin in different life stages [128, 132, 136, 137] due to *de novo* biosynthesis rather than sequestration as the larval host plants, in most cases, do not form aliphatic cyanogenic glucosides [4].

Among arthropods, millipedes also contain cyanogenic compounds, namely cyanohydrins such as mandelonitrile, and use them as defense against predators as discussed by Shear [133]. Synthesis of cyanide and cyanohydrins such as mandelonitrile was demonstrated for different species of millipedes using feeding tests with ^{14}C-phenylalanine and further radioactively labeled precursors [14, 107, 138]. As this resulted in labeling of phenylacetaldoxime and phenylacetonitrile as potential pathway intermediates, a biosynthesis pathway very alike the one described in plants and later in insects was proposed [14, 107, 138] (**Figure 4**). The unwanted release of cyanide is prevented by specifically shaped, two chamber glands where cyanide precursor and the hydrolyzing enzyme α-hydroxynitrile lyase are stored separately [107, 133]. In the α-hydroxynitrile chamber, organic acids generate low pH values to stabilize the α-hydroxynitrile [133]. Upon attack by a predator, gland secretions are mixed to generate cyanide. This mechanism allows the millipede to liberate cyanide in a controlled way, thereby economizing its chemical defense and protecting its own tissue from poisoning.

Although the presence of cyanogenic glucosides and, in part, their *de novo* biosynthesis has been described in selected species of the centipedes [139], millipedes and insects, and these

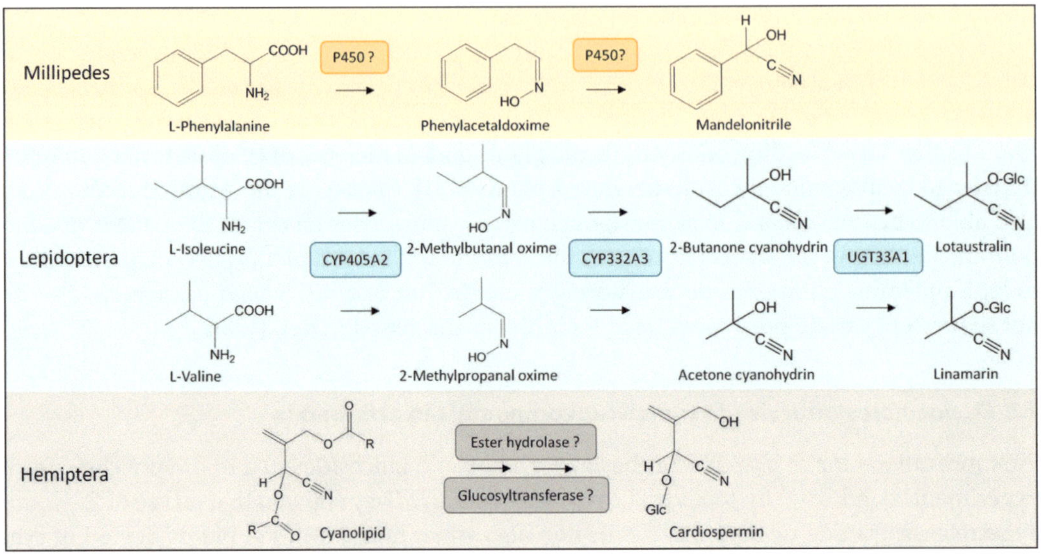

Figure 4. Biosynthesis of cyanogenic defense compounds in arthropods. Shown are the most widespread compounds and the enzymes characterized or proposed for the corresponding pathway. The biosynthesis of linamarin and lotaustralin in *Zygaena filipendulae* has been completely elucidated [16]. For millipedes, a similar reaction pathway is indicated by the reaction intermediates identified so far [14, 107, 138]. Cardiospermin is found in species of the Hemiptera, but its biosynthesis pathway is still unknown. Cyanolipids have been proposed as precursors [129].

phyla derive from a common ancestor, the absence of cyanogens in most insect species and the diversity of cyanogenic glucosides in combination with the long time span since their diversion (>390 MYA) [140] indicate that the biosynthetic pathways evolved independently in the different phyla [4]. Interestingly, insects biosynthesizing cyanogenic glucosides *de novo* also express both enzymes needed for their degradation and cyanide liberation, β-glucosidase and α-hydroxynitrile lyase [141].

7.3. Sequestration of cyanogenic compounds from the food plants

Most data known on sequestration of cyanogenic compounds in insects were generated using larvae of *Z. filipendulae* as a model. *Z. filipendulae* sequesters the cyanogenic glucosides linamarin and lotaustralin from its host plant *L. corniculatus* as intact glucosides, transports them into all tissues and retains them upon metamorphosis where they can be found in all tissues, too, as studies with isotopically labelled linamarin and lotaustralin have shown [142]. Interestingly, larvae of *Z. filipendulae* are also able to sequester the aromatic cyanogenic glucoside prunasin which they do not come in contact with naturally [118]. This indicates a versatile transport mechanism [118] and an evolutionary origin from a more general mechanism to deal with host plant xenobiotics.

Sequestration of cyanogens has also been shown for gynocardin, a cyclic α-hydroxynitrile glucoside, in *Acraea horta* (Lepidoptera:Nymphalidae) [143] and for sarmentosin, a β-hydroxynitrile, in *Parnassius phoebus* (Lepidoptera:Papilionidae) [144]. Although the latter compound is not cyanogenic *per se*, enzymatic catalysis may lead to the liberation of cyanide. Alternatively, a function of the intact compound as deterrent against predators is discussed [144].

Larvae of the bug *Leptocoris isolata* (Hemiptera:Serinethinae), but not the adult bugs, were shown to contain cyanogenic glucosides such as cardiospermin not present in their host plant [129]. However, their host plant produces cyanolipids containing the same aglucone. Therefore, it was proposed that this species recruited enzymes of xenobiotic metabolism to transform sequestered cyanolipids into cyanogenic glucosides for use in its own defense [129] (**Figure 4**).

7.4. Benefit of cyanogenesis for herbivores

Cyanogenic glucosides derived from *de novo* biosynthesis and sequestration may play a pivotal role in herbivore defense against predators as has been demonstrated with the model species *Z. filipendulae*. Larvae of *Z. filipendulae* react to aggression such as pricking by release of a defensive fluid from their dorsal cavities. This fluid contains, next to the cyanide detoxification product β-cyanoalanine, high amounts of the two aliphatic cyanogenic glucosides, linamarin and lotaustralin, and, among other proteins, their hydrolysis enzyme, β-glucosidase [100, 145]. Thus, cyanogenic glucosides and cyanide are used for deterrence and intoxication of potential predators. If whole larvae are ingested by a mammalian predator, another mechanism comes into effect. The larval hemolymph contains high amounts of β-glucosidase inactivated by a high pH value. Upon release of these proteins into the highly acidic environment of the predator's stomach, the β-glucosidase gets activated and cyanogenic glucoside hydrolysis leads to cyanide poisoning of the predator (discussed in Ref. [4]).

Nevertheless, cyanide alone is not always efficient for the animal's defense. It has been shown that in millipedes, not cyanide itself but the second product of mandelonitrile hydrolysis, benzaldehyde, is repellent to ants [49]. In contrast, for the intact cyanogenic glucoside cardiospermin a deterrent effect on ants has been shown which could not be observed for any cyanogenic glucoside before [129]. Thus, predating insects facing cyanide in their prey may have evolved a sensitive perception of substances with stronger odor or taste usually occurring alongside cyanide [49]. Alternatively, it was proposed that cyanide is the main means of defense against vertebrate predators, while benzaldehyde is used to repel arthropod enemies [133]. Phylogenetic and physiological data indicate that cyanogenesis as defense strategy has been lost and replaced by ancient phenolic defense compounds in some groups among the Polydesmida, mainly those unlikely to be targeted by vertebrate predators [133].

Next to their defensive roles, linamarin and lotaustralin are also used as nitrogen sources for chitin biosynthesis based on their turnover during metamorphosis [146]. During the formation of the pupal cuticle, cyanogenic glucosides are a key nitrogen source [146, 147]. However, mobilization through β-cyanoalanine synthase and nitrilase/nitrile hydratase leading to the formation of asparagine and aspartate similar to plants [148] has not been demonstrated in insects so far. The efficient transport of cyanogenic glucosides in *Z. filipendulae* is nevertheless evident based on the occurrence of these compounds in the wings, an organ devoid of any biosynthetic activity [146]. The important role of sequestered and *de novo*-biosynthesized cyanogenic glucosides for the defense and metabolism of the lepidopteran specialist *Z. filipendu-*

lae was additionally proven by the finding that male adult butterflies transfer the compounds to their partners as nuptial gift upon mating [142, 147].

Based on transcriptional and metabolite analyses [149], it has been hypothesized that the biosynthesis of cyanogenic glucosides in arthropods is older than their sequestration [149]. The biosynthesis is thought to have been constitutive in the ancestors of Zygaena which did not live on cyanogenic plants (but on Celastraceae) and were, therefore, not able to receive cyanogenic glucosides from their host plants. Upon exploration of cyanogenic plants, the insects' endogenous biosynthesis became inducible as sequestration helped to reduce metabolic costs for *de novo* biosynthesis [149]. Thus, the ability to handle and *de novo* biosynthesize cyanogenic glucosides allowed the moths to extend their host plant range and to even exploit the newly acquired host plants, cyanogenic glucoside-producing Fabaceae, to conserve energy and nutrients otherwise needed for the biosynthesis of these compounds [4, 128].

8. Conclusions and perspectives

The past 15 years have witnessed an enormous progress in our understanding of herbivore adaptational mechanisms to plant cyanide defenses and their evolution. A lot of the present knowledge has been acquired through the application of state-of-the-art analytical and molecular tools as well as imaging techniques to the model species *Z. filipendulae*, a well-known specialist on cyanide-defended plants which is able to sequester and *de novo* synthesize cyanogenic glucosides. Detailed studies of its feeding mode and gut physiology, the properties of its gut β-glucosidase and the evolution of its own biosynthetic pathway for cyanogenic glucosides have illuminated the role of cyanogenic glucosides as defenses and nitrogen storage compounds which have shaped coevolutionary relations between herbivores and their host plants. The availability of genome and transcriptome data of diverse animal species has fueled studies into cyanide detoxification mechanisms in herbivores. Together with modern analytical techniques, molecular biology has enabled identification of the first animal β-cyanoalanine synthases. As an interesting evolutionary background, these studies revealed that arthropod β-cyanoalanine synthases have likely been acquired through horizontal gene transfer from microbial gut symbionts. Proof of rhodanese activity of putative enzymes from arthropod databases still needs to be provided.

Future research will have to extend the present insights by studying a broader range of species with respect to their behavioral, physiological and biochemical adaptations to cyanogens. Besides the identification and detailed characterization of cyanide detoxification enzymes from additional species, transporters involved in cyanogen sequestration will be an interesting target of future investigations. In addition, experimental proof of essential roles of herbivore proteins involved in overcoming plant cyanide defenses might become possible *in vivo* through RNA interference or genome editing. Taken together, studies on herbivore adaptations to plant cyanide defenses are a prime example of the added value of multidisciplinary research combining ecology, physiology, biochemistry, systematics and genetics to provide insights into coevolution of herbivores and their food plants.

Acknowledgements

Financial support of our research on cyanide detoxification in arthropods by the German Research Foundation is gratefully acknowledged.

Author details

Maike van Ohlen[1,2]*, Anna-Maria Herfurth[1] and Ute Wittstock[1]

*Address all correspondence to: m.vanohlen@hotmail.com

1 Institute of Pharmaceutical Biology, Braunschweig University of Technology, Braunschweig, Germany

2 SeSaM Biotech GmbH, Aachen, Germany

References

[1] Hartmann T. From waste products to ecochemicals: Fifty years research on plant secondary metabolism. Phytochemistry. 2007;**68**(22–24):2831–2846.

[2] Fraenkel G.S. The Raison d'être of secondary plant substances – These odd chemicals arose as a means of protecting plants from insects and now guide insects to food. Science. 1959;**129**(3361):1466–1470.

[3] Raybuck S.A. Microbes and microbial enzymes for cyanide degradation. Biodegradation. 1992;**3**(1):3–18.

[4] Zagrobelny M., Bak S., Rasmussen A.V., Jørgensen A.B., Naumann C.M., Møller B.L. Cyanogenic glucosides and plant-insect interactions. Phytochemistry. 2004;**65**: 293–306.

[5] Møller B.J. Functional diversifications of cyanogenic glucosides. Curr. Opin. Plant Biol. 2010;**13**(3):337–346.

[6] Pičmanová M., Neilson E.H., Motawia M.S., Olsen C.E., Agerbirk N., Gray C.J., et al. A recycling pathway for cyanogenic glycosides evidenced by the comparative metabolic profiling in three cyanogenic plant species. Biochem. J. 2015;**469**(3):375–389.

[7] Selmar D., Lieberei R., Biehl B. Mobilization and utilization of cyanogenic glucosides. Plant Physiol. 1988;**86**:711–716.

[8] Kongsawadworakul P., Viboonjun U., Romruensukharom P., Chantuma P., Ruderman S., Chrestin H. The leaf, inner bark and latex cyanide potential of *Hevea brasiliensis*: Evidence for involvement of cyanogenic glucosides in rubber yield. Phytochemistry. 2009;**70**:730–739.

[9] Conn E.E. Cyanogenic compounds. Annu. Rev. Plant Physiol. 1980;**31**(1):433–451.

[10] Zagrobelny M., Bak S., Møller B.L. Cyanogenesis in plants and arthropods. Phytochemistry. 2008;**31**(1):433–451.

[11] Jørgensen K., Morant A.V., Morant M., Jensen N.B., Olsen C.E., Kannangara R., et al. Biosynthesis of the cyanogenic glucosides linamarin and lotaustralin in Cassava: Isolation, biochemical characterization, and expression pattern of CYP71E7, the oxime-metabolizing cytochrome P450 enzyme. Plant Physiol. 2011;**155**:282–292.

[12] Bak S., Paquette S.M., Morant M., Morant A.V., Saito S., Bjarnholt N., et al. Cyanogenic glucosides: A case study for evolution and application of cytochromes P450. Phytochem. Rev. 2006;**5**(2):309–329.

[13] Morant A.V., Bjarnholt N., Kragh M.E., Kjærgaard C.H., Jørgensen K., Paquette S.M., et al. The β-glucosidases responsible for bioactivation of hydroxynitrile glucosides in *Lotus japonicus*. Plant Physiol. 2008;**147**:1072–1091.

[14] Duffey S.S., Underhill E.W., Towers G.H.N. Intermediates in the biosynthesis of HCN and benzaldehyde by a polydesmid millipede, *Harpaphe haydeniana* (Wood). Comp. Biochem. Physiol. 1974;**47**(B):753–766.

[15] Duffey S.S. Cyanide and arthropods. In: Vennesland B., Conn E.E., Knowles C.J., Westley J., Wissing F., editors. Cyanide in Biology. London: Acad. Press; 1981. p. 385–414.

[16] Jensen N.B., Zagrobelny M., Hjernø K., Olsen C.E., Houghton-Larsen J., Borch J., et al. Convergent evolution in biosynthesis of cyanogenic defence compounds in plants and insects. Nat. Commun. 2011;**2**:273.

[17] Davis R.H., Nahrstedt A. Biosynthesis of cyanogenic glucosides in butterflies and moths – effective incorporation of 2-methylpropanenitrile and 2-methylbutanenitrile into linamarin and lotaustralin by Zygaena and Heliconius species (Lepidoptera). Insect Biochem. 1987;**17**:689–693.

[18] Zagrobelny M., Møller B.L. Cyanogenic glucosides in the biological warfare between plants and insects: The Burnet moth-Birdsfoot trefoil model system. Phytochemistry. 2011;**72**(13):1585–1592.

[19] Zagrobelny M., Scheibye-Alsing K., Jensen N., Møller B., Gorodkin J., Bak S. 454 Pyrosequencing based transcriptome analysis of *Zygaena filipendulae* with focus on genes involved in biosynthesis of cyanogenic glucosides. BMC Genomics. 2009; **10**(1):574.

[20] Mikolajczak K.L. Cyanolipids. Progr. Chem. Fats Other Lipids. 1977;**15**(2):97–130.

[21] Seigler D.S., Kawahara W. New reports of cyanolipids from Sapindaceaous plants. Biochem. Syst. Ecol. 1978;**4**:263–265.

[22] Seigler D.S., Mikolajczak K.L., Smith C.R. Jr., Wolff I.A., Bates R.B. Structure and reactions of a cyanogenetic lipid from *Cordia verbenacea* DC. seed oil. Chem. Phys. Lipids. 1970;**4**(2):147–161.

[23] Halkier B.A., Gershenzon J. Biology and biochemistry of glucosinolates. Annu. Rev. Plant Biol. 2006;**57**(1):303–333.

[24] Hopkins R.J., Van Dam N.M., Van Loon J.J.A. Role of glucosinolates in insect-plant relationships and multitrophic interactions. Annu. Rev. Entomol. 2009;**54**:57–83.

[25] Ettlinger M.G., Lundeen A.J. First synthesis of a mustard oil glucoside; the enzymatic Lossen rearrangement. J. Am. Chem. Soc. 1957;**79**(7):1764–1765.

[26] Wittstock U., Burow M. Tipping the scales – specifier proteins in glucosinolate hydrolysis. IUBMB Life. 2007;**59**(12):744–751.

[27] Kissen R., Bones A.M. Nitrile-specifier protein involved in glucosinolate hydrolysis in *Arabidopsis thaliana*. J. Biol. Chem. 2009;**284**(18):12057–12070.

[28] Burow M., Losansky A., Müller R., Plock A., Kliebenstein D.J., Wittstock U. The genetic basis of constitutive and herbivore-induced ESP-independent nitrile formation in Arabidopsis. Plant Physiol. 2009;**149**(1):561–574.

[29] Wittstock U., Kurzbach E., Herfurth A.-M., Stauber E.J. Glucosinolate breakdown. In: Kopriva S., editor. Advances in Botanical Research. Amsterdam: Elsevier 2016;80:125–169.

[30] Stauber E.J., Kuczka P., van Ohlen M., Vogt B., Janowitz T., Piotrowski M., Beuerle T., Wittstock U. Turning the 'mustard oil bomb' into a 'cyanide bomb': Aromatic glucosinolate metabolism in a specialist insect herbivore. PLoS One. 2012;**7**(4):e35545.

[31] Wittstock U., Agerbirk N., Stauber E.J., Olsen C.E., Hippler M., Mitchell-Olds T., et al. Successful herbivore attack due to metabolic diversion of a plant chemical defense. Proc. Natl. Acad. Sci. U. S. A. 2004;**101**(14):4859–4864.

[32] Wheat C.W., Vogel H., Wittstock U., Braby M.F., Underwood D., Mitchell-Olds T. The genetic basis of a plant-insect coevolutionary key innovation. Proc. Natl. Acad. Sci. U. S. A. 2007;**104**(51):20427–20431.

[33] Vergara F., Svatoš A., Schneider B., Reichelt M., Gershenzon J., Wittstock U. Glycine conjugates in a lepidopteran insect herbivore – The metabolism of benzylglucosinolate in the cabbage white butterfly, *Pieris rapae*. ChemBioChem. 2006;**7**(12):1982–1989.

[34] Vergara F., Svatoš A., Schneider B., Reichelt M., Gershenzon J., Wittstock U. Glycine conjugates in a lepidopteran insect herbivore – The metabolism of benzylglucosinolate in the cabbage white butterfly, *Pieris rapae*. ChemBioChem. 2007;**8**(15):1757.

[35] Jeschke V., Gershenzon J., Vassão D.G. Metabolism of glucosinolates and their hydrolysis products in insect herbivores. In: Jetter R., editor. Recent Advances Phytochemistry: The Formation, Structure and Activity of Phytochemicals. Switzerland: Springer International Publishing; 2015. p. 163–194.

[36] Frisch T., Motawia M.S., Olsen C.E., Agerbirk N., Møller B.L., Bjarnholt N. Diversified glucosinolate metabolism: Biosynthesis of hydrogen cyanide and of the hydroxynitrile glucoside alliarinoside in relation to sinigrin metabolism in *Alliaria petiolata*. Front. Plant Sci. 2015;**6**:926.

[37] Kuchernig J., Burow W., Wittstock U. Evolution of specifier proteins in glucosinolate-containing plants. BMC Evolutionary Biology. 2012;**12**:127.

[38] Ohkawa H., Ohkawa R., Yamamoto I., Casida J.E. Enzymatic mechanisms and toxicological significance of hydrogen cyanide liberation from various organothiocyanates and organonitriles in mice and houseflies. Pesticide Biochem. Physiol. 1972;**2**:95–112.

[39] Peiser G.D., Wang T.-T., Hoffman N.E., Yang S.F., Liu H., Walsh C.T. Formation of cyanide from carbon 1 of 1-aminocyclopropane-1-carboxylic acid during its conversion to ethylene. Proc. Natl. Acad. Sci. U.S.A. 1984;**81**(10):3059–3063.

[40] Yip W.-K., Yang S.F. Cyanide metabolism in relation to ethylene production in plant tissues. Plant Physiol. 1988;**88**(2):473–476.

[41] Schappert P.J., Shore J.S. Cyanogenesis, herbivory and plant defense in *Turnera ulmifolia* on Jamaica. EcoScience. 1999;**6**:511–520.

[42] Van Buuren K.J.H., Zuurendonk P.F., Van Gelder B.F., Muijsers A.O. Biochemical and biophysical studies on cytochrome aa3. V. Binding of cyanide to cytochrome aa3. Biochim. Biophys. Acta. 1972;**256**(2):243–257.

[43] Beasley D.M., Glass W.I. Cyanide poisoning: Pathophysiology and treatment recommendations. Occup. Med. 1998;**48**(7):427–431.

[44] Solomonson L.P. Cyanide as a metabolic inhibitor. In: Vennesland B., Conn E.E., Knowles C.J., Westley J., Wissing F., editors. Cyanide in Biology. London: Acad. Press; 1981. pp. 427–431.

[45] Way J.L. Cyanide intoxication and its mechanism of antagonism. Ann. Rev. Pharmacol. Toxicol. 1984;**24**:451–481.

[46] Borowitz J.L., Born G.S., Isom G.E. Potentiation of evoked adrenal catecholamine release by cyanide: Possible role of calcium. Toxicology. 1988;**50**:37–45.

[47] Ardelt B.K., Borowitz J.L., Isom G.E. Brain lipid peroxidation and antioxidant protectant mechanisms following acute cyanide intoxication. Toxicology. 1989;**56**:147–154.

[48] Page A.B., Lubatti O.F. Fumigation of insects. Ann. Rev. Entomol. 1963;**8**(1):239–264.

[49] Peterson S.C. Breakdown products of cyanogenesis. Naturwissenschaften. 1986;**73**(10): 627–628.

[50] Bond E.J. Effect of oxygen on cyanide poisoning in insects. Nature. 1962;**193** (4819): 1002–1003.

[51] Bond E.J. The influence of oxygen on the metabolism of *Sitophilus granaries* (L.) during cyanide poisoning. J. Insect Physiol. 1965;**11**(6):765–777.

[52] Chew M.Y., Boey C.G. Rhodanese of tapioca leaf. Phytochemistry. 1972;**11**(1):167–169.

[53] Long K.Y., Brattsten L.B. Is rhodanese important in the detoxification of the dietary cyanide in southern armyworm (*Spodoptera eridania cramer*) larvae? Insect Biochem. 1982;**12**(4):367–375.

[54] Oi S. Purification and some properties of *Trametes sanguinea* rhodanese. Agric. Biol. Chem. 1973;**37**(3):629–635.

[55] Okonji R.E., James I.E., Madu J.O., Fagbohunka B.S., Agboola F.K. Purification and characterization of rhodanese from the hepatopancreas of garden snail, *Limicolaria flammea*. Ife J. Sci. 2015;**17**(2):289–303.

[56] Sörbo B.H. Rhodanese. Methods Enzymol. 1955;**17**(2):289–303.

[57] Villarejo M., Westley J. Mechanism of rhodanese catalysis of thiosulfate-lipoate oxidation-reduction. J. Biol. Chem. 1963;**238**(12):4016–4020.

[58] Gafar A.T., Nelson A.A., Kayode A.F. Properties of rhodanese from the liver of tilapia, *Oreochromis niloticus*, in Asejire Lake, Nigeria. Afri. J. Biochem. Res. 2014;**8**(3):74–83.

[59] Lewis H.B. The chemistry and metabolism of the compounds of sulfur. Annu. Rev. Biochem. 1935;**4**(1):149–168.

[60] Jarabak R., Westley J. Human liver rhodanese. Nonlinear kinetic behavior in a double displacement mechanism. Biochemistry (Mosc.). 1974;**13**(16):3233–3236.

[61] Miller D.M., Delgado R., Chirgwin J.M., Hardies S.C., Horowitz P.M. Expression of cloned bovine adrenal rhodanese. J. Biol. Chem. 1991;**266**(8):4686–4691.

[62] Ploegman J.H., Drent G., Kalk K.H., Hol W.G.J., Heinrikson R.L., Keim P., et al. The covalent and tertiary structure of bovine liver rhodanese. Nature. 1978;**273**(5658):124–129.

[63] Cipollone R., Ascenzi P., Visca P. Common themes and variations in the rhodanese superfamily. IUBMB Life. 2007;**59**(2):51–59.

[64] Nagahara N., Okazaki T., Nishino T. Cytosolic mercaptopyruvate sulfurtransferase is evolutionarily related to mitochondrial rhodanese. J. Biol. Chem. 1995;**270**(27):16230–16235.

[65] Westley J. Cyanide and sulfane sulfur. In: Vennesland B., Conn E.E., Knowles C.J., Westley J., Wissing F., editors. Cyanide in Biology. London: Acad. Press; 1981. pp. 61–76.

[66] Schlesinger P., Westley J. An expanded mechanism for rhodanese catalysis. J. Biol. Chem. 1974;**249**(3):780–788.

[67] Jarabak R., Westley J. 3-Mercaptopyruvate sulfurtransferase: Rapid equilibrium-ordered mechanism with cyanide as the acceptor substrate. Biochemistry. 1980;**19**:900–904.

[68] Jarabak R., Westley J. Steady-state kinetics of 3-mercaptopyruvate sulfurtransferase from bovine kidney. Arch. Biochem. Biophys. 1978;**185**(2):458–465.

[69] Vachek H., Wood J.L. Purification and properties of mercaptopyruvate sulfur transferase of *Escherichia coli*. Biochim. Biophys. Acta Enzymol. 1972;**258**(1):133–146.

[70] de Duve C., Pressman B.C., Gianetto R., Wattiaux R., Appelmans F. Tissue fractionation studies. 6. Intracellular distribution patterns of enzymes in rat-liver tissue. Biochem. J. 1955;**60**(4):604–617.

[71] Dudek M., Frendo J., Koj A. Subcellular compartmentation of rhodanese and 3-mercaptopyruvatesulphurtransferase in the liver of some vertebrate species. Comp. Biochem. Physiol. B Biochem. 1980;**65**(2):383–386.

[72] Agboola F.K., Okonji R.E. Presence of rhodanese in the cytosolic fraction of the fruit bat (*Eidolon helvum*) liver. J. Biochem. Mol. Biol. 2004;**37**(3):275–281.

[73] Cerletti P. Seeking a better job for an under-employed enzyme: Rhodanese. Trends Biochem. Sci. 1986;**11**(9):369–372.

[74] Ola-Mudathir F.K., Madaugwu E.N. Effects of *Allium cepa* Linn. on rhodanese activities and half-life of cyanide in the blood. Int. J. Biol. Chem. Sci. 2015;**9**(2):1004–1012.

[75] Schulz V. Clinical pharmacokinetics of nitroprusside, cyanide, thiosulphate and thiocyanate. Clin. Pharmacokin. 1984;**9**(3):239–251.

[76] Porter D.W., Nealley E.W., Baskin S.I. *In vivo* detoxification of cyanide by cystathionase γ-lyase. Biochem. Pharmacol. 1996;**52**(6):941–944.

[77] Pagani S., Bonomi F., Cerletti P. Enzymic synthesis of the iron-sulfur cluster of spinach ferredoxin. Eur. J. Biochem. 1984;**142**(2):361–366.

[78] Clemedson C.J., Fredriksson T., Hansen B., Hultman H., Sörbo B. On the toxicity of sodium β-mercaptopyruvate and its antidotal effect against cyanide. Acta Physiol. Scand. 1958;**42**(1):41–45.

[79] Sörbo B. Enzymic transfer of sulfur from mercaptopyruvate to sulfite or sulfinates. Biochim. Biophys. Acta. 1957;**24**:324–329.

[80] Hatzfeld Y., Maruyama A., Schmidt A., Noji M., Ishizawa K., Saito K. β-Cyanoalanine synthase is a mitochondrial cysteine synthase-like protein in spinach and Arabidopsis. Plant Physiol. 2000;**123**(3):1163–1172.

[81] Ikegami F., Murakoshi I. Enzymic synthesis of non-protein β-substituted alanines and some higher homologues in plants. Phytochemistry. 1994;**35**(5):1089–1104.

[82] Kashiwamata S., Greenberg D.M. Studies on cystathionine synthase of rat liver properties of the highly purified enzyme. Biochim. Biophys. Acta Enzymol. 1970;**212**(3):488–500.

[83] Eliot A.C., Kirsch J.F. Pyridoxal phosphate enzymes: Mechanistic, structural, and evolutionary considerations. Annu. Rev. Biochem. 2004;**73**(1):383–415.

[84] Vozdek R., Hnízda A., Krijt J., Šerá L., Kožich V. Biochemical properties of nematode *O*-acetylserine(thiol)lyase paralogs imply their distinct roles in hydrogen sulfide homeostasis. Biochim. Biophys. Acta Proteins Proteomics. 2013;**1834**(12):2691–2701.

[85] Budde M.W., Roth M.B. The response of *Caenorhabditis elegans* to hydrogen sulfide and hydrogen cyanide. Genetics. 2011;**1075**(2):195–197.

[86] Meyers D.M., Ahmad S. Link between β-cyanoalanine synthase activity and differential cyanide sensitivity of insects. Biochim. Biophys. Acta Gen. Subj. 1991;**1075**(2):195–197.

[87] Wybouw N., Dermauw W., Tirry L., Stevens C., Grbic M., Feyereisen R., et al. A gene horizontally transferred from bacteria protects arthropods from host plant cyanide poisoning. eLife. 2014;**3**:e02365.

[88] Tai C.-H., Cook P.F. Pyridoxal 5′-phosphate-dependent α,β-elimination reactions: Mechanism of *O*-acetylserine sulfhydrylase. Acc. Chem. Res. 2001;**34**(1):49–59.

[89] Bonner E.R., Cahoon R.E., Knapke S.M., Jez J.M. Molecular basis of cysteine biosynthesis in plants – Structural and functional analysis of *O*-acetylserine sulfhydrylase from *Arabidopsis thaliana*. J. Biol. Chem. 2005;**280**(46):38803–38813.

[90] Yi H., Juergens M., Jez J.M. Structure of soybean β-cyanoalanine synthase and the molecular basis for cyanide detoxification in plants. Plant Cell. 2012;**24**(6):2696–2706.

[91] Bermúdez M.A., Páez-Ochoa M.A., Gotor C., Romero L.C. Arabidopsis S-sulfocysteine synthase activity is essential for chloroplast function and long-day light-dependent redox control. Plant Cell. 2010;**22**:403–416.

[92] Álvarez C., García I., Romero L.C., Gotor C. Mitochondrial sulfide detoxification requires a functional isoform *O*-acetylserine(thiol)lyase C in *Arabidopsis thaliana*. Mol. Plant. 2012;**5**(6):1217–1226.

[93] Wirtz M., Beard K.F.M., Lee C.P., Boltz A., Schwarzländer M., Fuchs C., et al. Mitochondrial cysteine synthase complex regulates *O*-acetylserine biosynthesis in plants. J. Biol. Chem. 2012;**287**(33):27941–27947.

[94] Wurtele E.S., Nikolau B.J., Conn E.E. Subcellular and developmental distribution of β-cyanoalanine synthase in barley leaves. Plant Physiol. 1985;**78**(2):285–290.

[95] van Ohlen M., Herfurth A.-M., Kerbstadt H., Wittstock U. Cyanide detoxification in an insect herbivore: Molecular identification of β-cyanoalanine synthases from *Pieris rapae*. Insect Biochem. Mol. Biol. 2016;**70**:99–110.

[96] Ressler C. Isolation and identification from Common Vetch of the neurotoxin β-cyano-L-alanine, a possible factor in neurolathyrism. J. Biol. Chem. 1962;**237**(3):733–735.

[97] Machingura M., Ebbs S.D. Functional redundancies in cyanide tolerance provided by β-cyanoalanine pathway genes in *Arabidopsis thaliana*. Int. J. Plant Sci. 2014;**175**(3): 346–358.

[98] Blumenthal-Goldschmidt S., Butler G.W., Conn E.E. Incorporation of hydrocyanic acid labelled with carbon-14 into asparagine in seedlings. Nature. 1963;**197**(4868):718–719.

[99] Kobayashi M., Shimizu S. Nitrile hydrolases. Curr. Opin. Chem. Biol. 2000;**4**(1):95–102.

[100] Pentzold S., Zagrobelny M., Khakimov B., Engelsen S.B., Clausen H., Petersen B.L., et al. Lepidopteran defense droplets – A composite physical and chemical weapon against potential predators. Scientific Reports. 2016;**6**:22407.

[101] Aminlari L., Tavana M., Kazemipour N. Rhodanese distribution in different tissues of rabbits. Comp. Clin. Pathol. 2014;**23**(5):1259–1262.

[102] Drawbaugh R.B. Marrs T.C. Interspecies differences in rhodanese (thiosulfate sulfurtransferase, EC 2.8.1.1) activity in liver, kidney and plasma. Comp. Biochem. 1987;**86**(2):307–310.

[103] Beesley S.G., Compton S.G., Jones D.A. Rhodanese in insects. J. Chem. Ecol. 1985;**11**(1):45–50.

[104] Ogunlabi O.O., Agboola F.K. A soluble β-cyanoalanine synthase from the gut of the variegated grasshopper *Zonocerus variegatus* (L.). Insect Biochem. Mol. Biol. 2007;**37**(1):72–79.

[105] Duffey S.S., Blum M.S. Phenol and guaiacol: Biosynthesis, detoxification, and function in a polydesmid millipede, *Oxidus gracilis*. Insect Biochem. 1977;**7**(1):57–65.

[106] Witthohn K., Naumann C.M. The distribution of β-cyanoalanine in cyanogenic Lepidoptera. Z. Naturforsch. 1984;**39**(c):837–840.

[107] Duffey S.S., Towers G.H.N. On the biochemical basis of HCN production in the millipede *Harpaphe haydeniana* (Xystodesmidae: Polydesmida). Can. J. Zool. 1978;**56**(1):7–16.

[108] Bond E.J. The action of fumigants on insects: III. The fate of hydrogen cyanide in *Sitophilus granarius* (L.). Can. J. Biochem. Physiol. 1961;**39**(12):1793–1802.

[109] Jones D.A., Parsons J., Rothschild M. Release of hydrocyanic acid from crushed tissues of all stages in the life-cycle of species of the Zygaeninae (Lepidoptera). Nature. 1962;**193**:52–53.

[110] Witthohn K., Naumann C.M. Cyanogenesis – a general phenomenon in the lepidoptera? J. Chem. Ecol. 1987;**13**(8):1789–1809.

[111] Bessie I.U., Agboola F.K. Detoxification of cyanide in insects. I. Purification and some properties of rhodanese from the gut of the variegated grasshopper *Zonocerus variegatus* (Orthoptera:Pyrgomorphidae). Int. J. Trop. Insect Sci. 2013;**33**(3):153–162.

[112] Sun B.F., Xiao J.H., He S.M., Liu L., Murphy R.W., Huang D.W. Multiple ancient horizontal gene transfer events and duplications in Lepidopteran species. Insect Mol. Biol. 2013;**22**(1):72–87.

[113] Crisp A., Boschetti C., Perry M., Tunnacliffe A., Micklem G. Expression of multiple horizontally acquired genes is a hallmark of both vertebrate and invertebrate genomes. Genome Biol. 2015;**16**:50.

[114] Wybouw N., Zhurov V., Martel C., Bruinsma K.A., Hendrickx F., Grbić V., et al. Adaptation of a polyphagous herbivore to a novel host plant extensively shapes the transcriptome of herbivore and host. Mol. Ecol. 2015;**24**:4647–4663.

http://dx.doi.org/10.5772/66277

[115] Ehrlich P.R., Raven P.H. Butterflies and plants: A study in coevolution. Evolution. 1964;**18**(4):586–608.

[116] Gleadow R.M., Woodrow I.E. Constraints on effectiveness of cyanogenic glycosides in herbivore defense. J. Chem. Ecol. 2002;**28**(7):1301–1313.

[117] Webber J.J., Roycroft C.R., Callinan J.D. Cyanide poisoning of goats from sugar gums (*Eucalyptus cladocalyx*). Austral. Vet. J. 1985;**62**:28.

[118] Pentzold S., Zagrobelny M., Bjarnholt N., Kroymann J., Vogel H., Olsen C.E., et al. Metabolism, excretion and avoidance of cyanogenic glucosides in insects with different feeding specialisations. Insect Biochem. Mol. Biol. 2015;**66**:119–128.

[119] Frehner M., Conn E.E. The linamarin β-glucosidase in Costa Rican wild lima beans (*Phaseolus lunatus* L.) is apoplastic. Plant Physiol. 1987;**84**:1296–1300.

[120] Pentzold S., Zagrobelny M., Roelsgaard P.S., Møller B.L., Bak S. The multiple strategies of an insect herbivore to overcome plant cyanogenic glucoside defence. PLoS One. 2014;**9**(3):e91337.

[121] Falk K.L., Gershenzon J. The desert locust, *Schistocerca gregaria*, detoxifies the glucosinolates of *Schouwia purpurea* by desulfation. J. Chem. Ecol. 2007;**33**(8):1542–1555.

[122] Ballhorn D.J., Lieberei R., Ganzhorn J.U. Plant cyanogenesis of *Phaseolus lunatus* and its relevance for herbivore-plant interaction: The importance of quantitative data. J. Chem. Ecol. 2005;**31**(7):1445–1473.

[123] Ferreira C., Parra J.R.P., Terra W.R. The effect of dietary plant glycosides on larval midgut β-glucosidases from *Spodoptera frugiperda* and *Diatraea saccharalis*. Insect Biochem. Mol. Biol. 1997;**27**(1):55–59.

[124] Engler H.S., Spencer K.C., Gilbert L.E. Preventing cyanide release from leaves. Nature. 2000;**406**:144–145.

[125] Hall F.R., Hollingworth R.M., Shankland D.L. Cyanide tolerance in Millipedes: The biochemical basis. Comp. Biochem. Physiol. 1971;**38**(B):723–737.

[126] Heisler C.R., Hodnick W.F., Ahmad S. Evidence for target site insensitivity to cyanide in *Spodoptera eridania* larvae. Comp. Biochem. Physiol. Part C Comp. Pharmacol. 1988;**91**(2):469–472.

[127] Hammer T.J., Bowers M.D. Gut microbes may facilitate insect herbivory of chemically defended plants. Oecologia. 2015;**179**:1–4.

[128] Nahrstedt A. Cyanogenesis and the role of cyanogenic compounds in insects. Ciba Found. Symp. 1988;**140**:131–150.

[129] Braekman J.C., Daloze T.D., Pasteels J.M. Cyanogenic and other glucosides in a Neo-Guinean bug *Leptocoris isolata*: Possible precursors in its host-plant. Biochem. Syst. Ecol. 1982;**10**(4):355–364.

[130] Davis R.H., Nahrstedt A. Linamarin and lotaustralin as the source of cyanide in *Zygaena filipendulae* L. (Lepidoptera). Comp. Biochem. Physiol. 1979;**64**(B):395–397.

[131] Schappert P.J., Shore J.S. Effects of cyanogenesis polymorphism in *Turnera ulmifolia* on *Euptoieta hegesia* and potential Anolis predators. J. Chem. Ecol. 1999;**25**(6):1455–1479.

[132] Nahrstedt A., Davis R.H. Occurrence, variation and biosynthesis of the cyanogenic glucosides linamarin and lotaustralin in species of the Heliconiini (Insecta, Lepidoptera). Comp. Biochem. Physiol. B Biochem. Mol. Biol. 1983;**75**:65–73.

[133] Shear W.A. The chemical defenses of millipedes (Diplopoda): Biochemistry, physiology and ecology. Biochem. System. Ecol. 2015;**61**:78–117.

[134] Wray V., Davis R.H., Nahrstedt A. Biosynthesis of cyanogenic glucosides in butterflies and moths: Incorporation of valine and isovaline into linamarin and lotaustralin by Zygaena and Heliconius species (Lepidoptera). Z. Naturforsch. 1983;**38**(c):583–588.

[135] Holzkamp G., Nahrstedt A. Biosynthesis of cyanogenic glucosides in the Lepidoptera – incorporation of [U-C-14]-2-methylpropanealdoxime, 2S-[U-C-14]- methylbutanealdoxime and D,L-[U-C-14]-N-hydroxyisoleucine into linamarin and lotaustralin by the larvae of *Zygaena trifolii*. Insect Biochem. Mol. Biol. 1994;**24**:161–165.

[136] Brown K.S., Francini R.B. Evolutionary strategies of chemical defense in aposematic butterflies: Cyanogenesis in Asteraceae-feeding American Acreinae. Chemoecology. 1990;**1**:52–56.

[137] Nahrstedt A., Davis R.H. The occurrence of the cyanoglucosides, linamarin and lotaustralin, in Acraea and Heliconius butterflies. Comp. Biochem. Physiol. B Biochem. Mol. Biol. 1981;**68**:575–577.

[138] Towers G.H.N., Duffey S.S., Siegel S.M. Defensive secretion: Biosynthesis of hydrogen cyanide and benzaldehyde from phenylalanine by a millipede. Can. J. Zool. 1972;**50**(7):1047–1050.

[139] Jones H.T., Conner W.E., Meinwald J., Eisner H.E., Eisner T. Benzoyl cyanide and mandelonitrile in the cyanogenetic secretion of a centipede. J. Chem. Ecol. 1976;**2**(4):421–429.

[140] Gaunt M.W., Miles M.A. An insect molecular clock dates the origin of the insects and accords with palaeontological and biogeographic landmarks. Mol. Biol. Evol. 2002;**19**:748–761.

[141] Müller E., Nahrstedt A. Purification and characterization of an α-hydroxynitrile lyase from the haemolymph of the larvae of *Zygaena trifolii*. Planta Med. 1990;**56**:611–612.

[142] Zagrobelny M., Olsen C.E., Pentzold S., Fürstenberg-Hägg J., Jørgensen K., Bak S., et al. Sequestration, tissue disruption and developmental transmission of cyanogenic glucosides in a specialist insect herbivore. Insect Biochem. Mol. Biol. 2014;**44**:44–53.

[143] Raubenheimer D. Cyanoglycoside gynocardin from *Acraea horta* (L.) (Lepidoptera:Acraeinae) – Possible implications for evolution of Acraeinae host choice. J. Chem. Ecol. 1989;**15**(8):2177–2189.

[144] Nishida R., Rothschild M. A cyanoglucoside stored by a Sedum-feeding Apollo butterfly, *Parnassius phoebus*. Experientia. 1995;**51**:267–269.

[145] Witthohn K., Naumann C.M. Qualitative and quantitative studies on the compounds of the larval defensive secretion of *Zygaena trifolii* (Esper, 1783) (Insecta, Lepidoptera, Zygaenidae). Comp. Biochem. Physiol. Part C Comp. Pharmacol. 1984;**79**(1):103–106.

[146] Fürstenberg-Hägg J., Zagrobelny M., Olsen C.E., Jørgensen K., Møller B.L., Bak S. Transcriptional regulation of *de novo* biosynthesis of cyanogenic glucosides throughout the life-cycle of the burnet moth *Zygaena filipendulae* (Lepidoptera). Insect Biochem. Mol. Biol. 2014;**49**:80–89.

[147] Zagrobelny M., Olsen C.E., Bak S., Møller B.L. Intimate roles for cyanogenic glucosides in the life cycle of *Zygaena filipendulae* (Lepidoptera, Zygaenidae). Insect Biochem. Mol. Biol. 2007;**37**:1189–1197.

[148] Lieberei R., Selmar D., Biehl B. Metabolization of cyanogenic glucosides in *Hevea brasiliensis*. Plant Syst. Evol. 1985;**150**:49–63.

[149] Fürstenberg-Hägg J., Zagrobelny M., Jørgensen K., Vogel H., Møller B.L., Bak S. Chemical defense balanced by sequestration and *de novo* biosynthesis in a Lepidopteran specialist. PLoS One. 2014;**9**(10):e108745.

Section 2

Herbivory and Food Processing of Grazing Animals

Chapter 3

Utilization of Biomarkers to Study the Grazing Behavior of Herbivore Species

Luis Miguel Mendes Ferreira, Rafael Celaya,
Ana Sofia Santos, Koldo Osoro and
Miguel Rodrigues

Additional information is available at the end of the chapter

http://dx.doi.org/10.5772/67345

Abstract

Knowledge on diet selection of different herbivore species under each specific vegetation community is essential to develop and apply appropriate management decisions for each grazing system in order to, simultaneously, have a more efficient and sustainable utilization of pasture resources and the best animal performance level. In this chapter, traditional and more recent methodologies that can be used for studying diet selection of both domestic and wild herbivores are briefly presented, identifying the main advantages and limitations of their use. Particular emphasis is given to the utilization of epicuticular compounds, namely alkanes, long-chain fatty acids and long-chain alcohols, as faecal markers. The validation of their use is presented taking into account studies performed with different animal species under controlled conditions. The main advantages and shortcomings for their application to field studies with grazing animals are highlighted. Data indicate that the combination of these epicuticular compounds seems promising to overcome the enumerated constraints, allowing its application to more complex vegetation communities.

Keywords: diet selection, faecal markers, ruminant species

1. Introduction

The success of the strategies for the management of herbivores grazing on different plant communities, driven by production or environmental goals, requires the understanding of the processes involved in plant-herbivore interactions and their consequences for both plants and herbivores [1]. The plant-herbivore interaction is mutual and dynamic. The structure, composition, productivity, nutritive value and distribution of the different plant communities determine the intake and nutritional status of the animals [2, 3]. In turn, the herbivores, through grazing, trampling, defecation, urination, etc., affect the dynamics of the vegetation community [4, 5]. These interrelationships are specific for each herbivore species and each vegetation type and are still poorly understood, leading to the use of less appropriate management strategies for agricultural and other land use objectives [4, 6]. The type (cutting, grazing or a mixed system) and intensity level of management will have a determinant role on the evolution of the habitat and on the biodiversity, being extremely important on the maintenance of species balance, maturity and nutritive value in plant communities, relating the timing and severity of defoliation in relation to patterns of plant growth and maturity, and proposed objectives (animal performance, biodiversity, sustainability, etc.).

The understanding of the grazing behaviour, especially diet selection, of different animal species under diverse conditions is essential to develop an appropriate grazing strategy for each specific situation in order to have a more efficient and sustainable utilization of the existing vegetation (**Figure 1**). The different dietary choices between plant species and plant parts in a specific vegetation community offered to the grazing animals are the main mechanism through which herbivores could increase sward heterogeneity [3, 7]. The diet selected by animals is constrained by temporal and spatial changes in the sward structure, plant defence mechanisms, food availability, plant phenology and animal factors [6, 7], and it differs between animal species [6, 8, 9] and also between breeds of cattle [10], sheep [11] and goats [12].

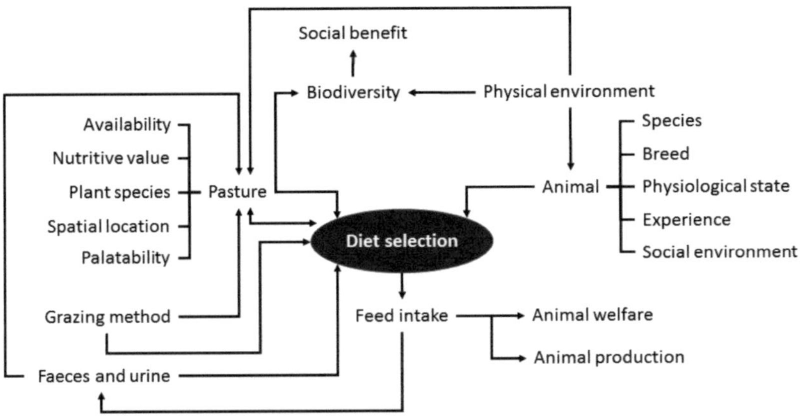

Figure 1. Schematic diagram of factors affecting diet selection of herbivores and its effects on animal production and biodiversity in a grazing system.

Generally, ruminant species are classified into three feeding types according to morphological and physiological adaptations of the digestive system [6, 13–15]: concentrate selectors (browsers), intermediate feeders and grass-roughage eaters (grazers). Based on this classification, it has been assumed that ruminant grazers, with greater body weight, achieve a higher extraction of nutrients from the diet consumed than browsers with low body weight [16]. According to Pérez-Barbería et al. [15] and Udén and van Soest [17], this is due to a higher extent of digestion of fibre by means of higher food retention in the rumen, larger stomach capacity, higher degree of stomach compartmentalization and smaller openings between the rumen and omasum. In contrast, small ruminants would compensate this lower digestion capacity by selecting high-quality plant parts such as fruits, pods, young shoots and leaves.

Previous studies [9, 18], carried out in heathland vegetation communities with adjacent areas of improved pasture (*Lolium perenne* and *Trifolium repens*) (**Figure 2**) across the grazing season (May–December), indicate an almost total preference for herbaceous species by cattle, contrasting with the higher preference for the woody species (*Erica* spp., *Calluna vulgaris* and *Ulex gallii*) revealed by goats. By contrast, horses and sheep showed an intermediate behaviour, increasing the selection of the woody species through the grazing season as a result of the decrease in the availability of the preferred herbaceous species [18]. These differences in the grazing behaviour were reflected in animal performance [19]. This distinct behaviour and variable responses of different animal species allow alternative strategies to develop viable systems aiming to achieve production and biodiversity outcomes. Therefore, the evaluation of diet composition of grazing animals is important for the achievement of sustainable management and production systems for each vegetation community.

Figure 2. Different herbivore species grazing heathland vegetation communities with adjacent areas of improved pasture of the north of Spain.

In this review, we aim to describe several methodologies that are available to assess plant-animal interactions, with particular relevance to the utilization of epicuticular compounds. The main advantages and limitations of each method are also explored, comparing the accuracy of diet composition estimates.

2. Techniques used to estimate diet composition in herbivores

Traditional techniques used to estimate diet composition of grazing animals are based either on measurements on the plant biomass (the utilization techniques) or on animal-based measurements [20], namely the direct observation of the grazing animal and the microhistological examination of plant fragments in different samples. However, all these techniques have important limitations associated with the measurement processes themselves, as the normal foraging behaviour may be compromised, and with the accuracy of the estimations [21].

Direct observation of the number of bites and the feeding times spent by the grazing animals on different plant communities is frequently used. The simplicity and minor equipment requirements are pointed out as advantages of this approach. However, as stated by Holechek et al. [20], it is extremely difficult to identify the plant species being consumed, especially when there is no spatial separation between plant species, and to convert the grazing times or number of bites to an accurate estimate of the amount of the plant consumed [22], besides being a time-consuming approach that is very difficult to accomplish during nocturnal periods.

The microhistological procedures rely on the visual identification of epidermal cuticular fragments in samples of oesophageal extrusa, in a gut compartment or in faeces [20, 21]. Diet composition is expressed in terms of the proportion of identifiable fragments coming from each plant species. Although microhistological approaches can be valuable to confirm the presence or the absence of a particular plant species or plant part in the diet [23], they are tedious to perform, require a lot of training of the researchers and involve sacrifice (stomach analysis) and fistulation (oesophageal extrusa) of the animals, unless faecal samples are used. Moreover, in the case of using faecal samples, possible differential digestion of the different plant species and the large proportion of unidentifiable fragments reduce the accuracy of diet composition estimates.

Another methodology that has been used for studying diet selection of herbivores is the near-infrared reflectance spectroscopy of faeces (F.NIRS) [24–27]. This methodology involves the association between faecal spectra with that of diets consumed, i.e. measurements of the reflectance of light between 700 and 2500 nm (for more details, see Dixon and Coates [28]). This spectrum gives a specific signature depending on the presence, character and number of important chemical bonds, such as OH, NH and CH [28]. According to Swain and Friend [29], one of the major limitations pointed out to NIRS applications (i.e. estimation of feed intake, digestibility and diet composition) is the need to have accurate calibration equations based on known and estimated nutritional parameters that will obviously vary for each specific situation (vegetation community). Nevertheless, these authors recognized the usefulness of this technique in identifying the presence of a specific feed item.

Results obtained by Ferreira et al. [18] suggest large variation in the spatial choice (i.e. plant communities where to graze) between animal species within a day and throughout the grazing season. The nutritive value, availability and the spatial distribution of the feed resources, and the distance to water and slope are major factors influencing grazing distribution patterns [30]. Early studies used visual field observations to assess these temporal and spatial modifications of rangeland use by both domestic and wild herbivores [30]. The utilization of recent available telemetry techniques can help grazing scientists to assess landscape vegetation preferences of herbivores [29], increasing the number of observed animals and reducing significantly the labour and allowing the collection of high-quality and unbiased data over a 24-h period. Identification of the preferred grazing sites can be accomplished by using telemetry devices, as global positioning systems (GPS). These devices are able to fine-scale spatio-temporal location data [31] with a spatial accuracy of <5 m [32] depending on the telemetry devices. This information together with data on the spatial arrangement of the plant communities can be used to assess the animals' patch selection. According to Swain and Friend [29], the spatial arrangement of vegetation (number and size of the patches) will determine the level of local accuracy needed, i.e. small patches in larger number will need a higher accuracy of location data. In a recent study, Thompson et al. [33] used GPS collars to spatially register cattle location, and based on this, data were able to assess their activities (grazing, travelling or resting) on distinct plant communities of a rangeland, using an algorithm developed to classify cattle activity. Hebblewhite and Haydon [31] referred that the high cost of GPS collars that depends on its features (i.e. battery size, longevity, programmability, remote data access) has led researchers to opt for using fewer GPS units, limiting statistical inference. According to these authors, collar failures that could range from 5 to 50% of the units reduce even further sample size. In addition to these shortcomings, this method does not allow to quantify or estimate diet composition.

Analysis of stable carbon isotopes in animal faeces has also been used to discriminate C_3 and C_4 plants on the diet selected by domestic [31–37] and wild herbivores [38]. This methodology is based on differences between plants with different photosynthesis pathways in fractioning of ^{13}C, with C_3 plants discriminating more against the heavier isotope ^{13}C in favour of ^{12}C than C_4 plants. This results in different ^{13}C:^{12}C ratios that are expressed as $\delta^{13}C$ relative to the ^{13}C:^{12}C ratio of the international Vienna Pee Dee Belemnite standard. Using these markers De Smet et al. [39] were able to estimate accurately the proportion of C_4 plant material in the diet analysing stable carbon isotope ratios ($\delta^{13}C$ value) in different tissues (blood, plasma, liver, kidney fat, hair, muscle and ruminal contents) taken from beef animals at slaughter. Nevertheless, Dove and Mayes [21] pointed out some limitations to this technique: (1) limited to situations where C_4 plants are present, for example, tropical grazing systems; (2) when using faecal samples, differential recovery of feeds in faeces may lead to underestimation of those of higher digestibility; and (3) possible effect of faecal endogenous carbon on the faecal carbon isotope ratio.

Alternatively, plant-wax components, especially alkanes and other wax components, such as long-chain alcohols and long-chain fatty acids, have been suggested as possible markers to estimate diet composition. The main advantages of using these markers is the fact that for their quantification the same analytical procedure is used on samples of the diet components and animal faeces, reducing labour and analytical error. Moreover, it provides the necessary information for the estimation of diet composition, digestibility and intake for each individual, therefore accommodating possible differences between individuals [21].

3. Epicuticular compounds

The aerial surfaces of most higher plants are covered by a layer of (epicuticular) wax that is a complex mixture of hydrophobic compounds such as long-chain fatty acids, aldehydes, alcohols, triterpenes, sterols, ketones, esters, flavonoids and alkanes [40]. According to Dove and Mayes [41], the chemical composition of this layer varies within plant species and plant parts, with leaves and floral parts tending to present higher wax concentrations than stems [21]. This layer has multiple functions, being the first line of protection between plants and the environment, acting as hydrophobic barriers, limiting nonstomatal water loss, and may constitute a defence mechanism against bacterial and fungal pathogens and other stress agents [40]. According to Eigenbrode and Espelie [42], it also plays an important role in the plant-insect interactions, repelling or attracting them.

Although the first studies on the possible use of epicuticular compounds as faecal markers were carried out with long-chain fatty acids (LCFAs) by Body and Hansen [43] and Grace and Body [44], alkanes are the ones most widely studied and applied in field studies due to their relative inertness and simplicity of analysis [21]. Alkanes present in the epicuticular mixture differ in carbon-chain length, varying from 21 to 37 carbon atoms [45]; those with odd number of carbon atoms represent more than 90% of the total content. Generally, the most abundant are the n-nonacosane (C_{29}), n-untriacontane (C_{31}) and n-tritriacontane (C_{33}) [22, 45]. The alkanes with less than 25 and more than 35 carbon atoms are present in very low concentrations. The alkane content varies between plant species (**Table 1**), plant parts and even cultivars of the same species [46, 47], plant stages of maturity and climatic conditions. In general, most of the herbaceous species, especially tropical forage species [48, 49], but also some shrub species (e.g. *U. gallii*, [50]), are known to possess very low alkane concentrations.

As can be observed in **Table 1**, differences in the alkane profiles between plant species occur in terms of absolute concentrations and relative proportions of the individual alkanes in the total content. Dove et al. [47] studied the effect of the plant species, age and part of the plant on the alkane profiles of different pasture species (*Phalaris aquatica, L. perenne, T. repens, Trifolium subterraneum* subsp. *subterraneum*, *T. subterraneum* subsp. *yanninicum* and *Medicago sativa*) and observed that species explained 85% and date of harvest only 6% of total variation. Differences in the alkane content between plant parts in the same pasture species were observed by Dove et al. [47]. Higher concentrations were found in the leaf than in the stem fraction. Also, much higher concentrations presented by the inflorescence of the perennial ryegrass (*L. perenne*) and by the flower of the white clover (*T. repens*) should be pointed out [47]. Less evident is the effect of age/stage of development in the alkane content of plant species. As stated above, the influence of the harvest date on the alkane content of the plant species studied by Dove et al. [47] accounted for only 5.7% of total variance. The results obtained by Oliveira et al. [57] indicate a decrease with age in the concentrations of C_{33} and C_{35} of hays of *Pennisetum purpureum* (C_{33} r = −0.97; C_{35} r = −0.99). Similar results were obtained by Laredo et al. [48] in leaves of *Pennisetum glaucum* (C_{33} r = −0.81; C_{35} r = −0.85) and *Sorghum* sp. (C_{33} r = −0.96; C_{35} r = −0.95) as age increased. However, opposite results were obtained by Smith et al. [58] when evaluating the effect of season of harvest on alkane concentrations of 40 common rangeland grasses found in Southern Africa. These authors did not observe a significant change in alkane concentrations either in leaf or stem components of the plant species between dry and wet seasons, suggesting that differences in alkane concentrations in whole plant samples could result from differences in the proportions of plant parts that present different alkane patterns.

Species	n-Alkanes (mg/kg DM)									References
	C_{25}	C_{26}	C_{27}	C_{28}	C_{29}	C_{30}	C_{31}	C_{32}	C_{33}	
Lolium perenne	20.0	5.2	40.2	12.7	178.0	18.8	274.0	12.0	115.4	[50]
Lolium multiflorum	33.5	5.4	56.6	8.6	150.8	13.1	181.6	5.1	23.7	[51]
Lolium rigidum	17.2	7.6	51.0	17.7	254.0	22.9	411.0	–	7.6	[52]
Festuca arundinacea	23.6	3.2	42.3	7.6	129.3	12.1	215.7	6.8	58.7	[53]
Holcus lanatus	133.6	15.8	111.5	21.8	225.3	23.8	178.0	12.0	47.2	[51]
Phalaris aquatica	10.7	4.2	8.4	3.8	14.2	3.8	22.2	–	7.6	[52]
Nardus stricta	19.9	5.0	73.1	18.1	535.9	26.3	647.9	17.5	243.5	[54]
Leymus chinensis	3.0	2.0	11.0	4.0	26.0	4.0	57.0	2.0	15.0	[55]
Leymus dasystachys	10.0	4.0	28.0	4.0	47.0	4.0	46.0	2.0	12.0	[55]
Elymus sibiricum	8.0	2.0	16.0	4.0	114.0	7.0	185.0	4.0	25.0	[55]
Trifolium repens	16.4	3.8	38.2	11.3	170.0	16.6	206.9	7.3	22.2	[50]
Trifolium striatum	10.0	4.0	48.2	30.0	989.9	22.5	68.1	5.1	7.9	[53]
Trifolium arvensis	30.2	9.2	122.7	33.9	915.2	40.7	314.2	20.9	32.5	[51]
Trifolium subterraneum	4.9	4.7	52.3	18.1	361.0	10.9	80.8	–	6.0	[52]
Vicia sativa	15.8	3.7	67.0	12.3	204.2	16.3	502.8	15.1	29.7	[51]
Ornithopus compressus	17.4	5.3	40.3	9.7	570.1	8.4	60.8	1.1	10.7	[51]
Ulex gallii	4.7	3.1	38.5	11.1	111.2	18.5	269.5	7.3	9.6	[50]
Calluna vulgaris	15.8	9.9	75.4	26.6	289.9	35.1	939.8	58.8	685.5	[54]
Erica cinerea	18.0	7.1	45.1	12.9	215.6	38.7	1196.7	59.4	493.4	[54]
Erica umbellata	17.6	7.7	51.2	12.9	239.7	30.9	580.6	35.9	235.7	[54]
Erica arborea	14.5	4.6	67.8	21.4	408.7	93.7	1625.8	133.9	662.2	[50]
Erica tetralix	7.0	5.0	50.0	18.0	926.0	45.0	1838.0	46.0	687.0	[56]
Vaccinium myrtillus	13.0	10.0	45.0	42.0	151.0	33.0	201.0	11.0	46.0	[56]

Table 1. Alkane concentrations (mg/kg DM) of several herbaceous and shrub plant species.

Other epicuticular compounds, namely long-chain fatty alcohols (LCFAs) [59–63] and LCOH [53, 64–66], have also been suggested as possible diet composition markers. Also, alkenes (unsaturated aliphatic hydrocarbons) were tested with success by Dove and Oliván [67] to estimate diet composition of sheep fed with different proportions of chaffed perennial ryegrass and unpelleted sunflower meal labelled with beeswax. These epicuticular compounds have the advantage over any other possible markers as the separation and quantification of these wax components can be an extension of the alkane procedure, not adding much more analytical work [59].

It should be noted that, as stated by Dove and Mayes [21], all studies have been based on total LCFA and LCOH concentrations (i.e. free plus esterified LCFA and LCOH), as a result of the cleavage of wax esters promoted by the saponification of samples with ethanolic KOH (1 M) in the extraction process. The LCFAs present in the epicuticular waxes are mainly mixtures of straight-chain saturated compounds [41] with an even number of carbons (**Table 2**). Within the LCFA that can be detected in animal faeces, those with carbon-chain lengths between C_{22} and C_{34} are suitable for diet composition estimation as they are exclusively associated with plant epicuticular waxes and present high recovery in animal faeces [41, 60]. Various studies have shown clear differences in the LCFA profiles between different plant species [41, 60–63, 68, 69], making them useful as diet composition markers. In general, individual and total LCFA concentrations of plant species are much higher than those found for the alkanes, especially for the herbaceous species [60, 62]. In fact, Ferreira et al. [60] and Lin et al. [69] observed that the majority of LCFAs with even-chain length in herbaceous species have concentrations above 100 mg/kg DM, whereas only a few alkanes exceeded this value. Also, Ali et al. [68] and Lin et al. [69] found total LCFA concentrations that were in average 10 times greater than the total alkane concentrations of 25 different rangeland species from Sudan and native Chinese grass species (*Leymus chinensis*, *Leymus dasystachys* and *Elymus sibiricum*), respectively. As also found for alkanes, differences between plant parts can also be observed in their LCFA profiles. Although there is limited information on possible differences between plant parts in their LCFA profiles, results obtained by Ferreira et al. [60] indicated a trend for the leaf/stem fraction of *L. perenne* to present higher concentrations on the longer (>25 carbon atoms) LCFA than the inflorescence fraction.

Species	**Even-chain fatty acids (mg/kg DM)**								**References**
	C_{20}-acid	C_{22}-acid	C_{24}-acid	C_{26}-acid	C_{28}-acid	C_{30}-acid	C_{32}-acid	C_{34}-acid	
Lolium perenne	–	514.3	381.9	559.2	396.3	287.1	128.2	32.2	[60]
Leymus chinensis	212.0	247.0	322.0	91.0	159.0	152.0	144.0	–	[69]
Leymus dasystachys	207.0	346.0	229.0	171.0	338.0	191.0	55.0	–	[69]
Elymus sibiricum	196.0	257.0	217.0	169.0	165.0	114.0	55.0	–	[69]
Trifolium repens	–	612.9	715.8	607.9	792.4	440.4	64.7	1.2	[60]
Ulex gallii	–	447.2	308.9	128.0	114.1	133.9	25.5	0.2	[60]
Agrostis-Poa[1]	–	324.2	249.5	522.6	195.9	156.7	97.6	42.3	[63]
Poa spp.[2]	–	236.1	156.7	366.6	120.1	45.3	24.0	6.5	[63]
Heather[3]	–	549.9	485.6	482.7	550.0	432.8	394.3	105.7	[63]
Calluna vulgaris	148.0	347.0	255.0	195.0	199.0	168.0	6.0	31.0	[68]
Erica arborea	–	645.2	292.7	215.0	434.1	782.9	528.7	138.5	[60]
Vaccinium myrtillus	211.0	179.0	140.0	128.0	325.0	1132.0	176.0	6.0	[68]

[1]Leaf fractions.

[2]Flowerstem fractions.

[3]Composed of *Erica umbellata* (0.76), *Erica cinerea* (0.16) and *Calluna vulgaris* (0.08).

Table 2. Even-chain fatty acid concentrations (mg/kg DM) of several herbaceous and shrub plant species.

Similarly to the LCFA, free LCOHs found in epicuticular wax of plant species are straight-chain saturated compounds with an even number of carbons within the same range of carbon-chain length referred to the LCFA (C_{20}–C_{34}) (**Table 3**). They are mainly primary alcohols, although many conifers present high concentrations of the odd-chain secondary alcohol 10-nonacosanol (C_{29}) [41]. As observed for the other epicuticular markers, LCOH profiles vary among plant species [52, 64, 65, 68, 69]. Generally, grass species are characterized by very high concentrations in C_{26} and C_{28} alcohols [52, 53, 64, 68], whilst C_{30} alcohol can be detected in large amounts in legumes [52, 64]. In general, total LCOH concentrations are within those of alkanes and LCFA, although Lin et al. [55, 69] reported a predominance of LCOH over the LCFA in *L. chinensis, L. dasystachys, E. sibiricum, Stipa baicalensis, Stipa grandis* and *Cleistogenes squarrosa*. As also found for the alkanes and LCFA, results suggest clear differences between vegetative and reproductive parts of herbaceous species. In fact, Ferreira et al. [64] indicated that the reproductive parts of *L. perenne* are characterized by having higher proportions of shorter LCOH than the vegetative tissues.

Species	**Even-chain alcohols (mg/kg DM)**							**References**
	1-C_{20}-ol	**1-C_{22}-ol**	**1-C_{24}-ol**	**1-C_{26}-ol**	**1-C_{28}-ol**	**1-C_{30}-ol**	**1-C_{32}-ol**	
Lolium perenne	47.3	61.5	162.5	2159.6	517.4	149.0	54.8	[64]
Lolium rigidum	–	12.3	60.0	1751.0	363.5	176.1	–	[52]
Festuca arundinacea	–	–	26.8	638.8	100.5	58.0	–	[53]
Phalaris aquatica	–	14.6	37.8	2813.0	134.9	65.3	–	[52]
Leymus chinensis	0	21.0	73.0	361.0	846.0	252.0	116.0	[69]
Leymus dasystachys	9.0	28.0	142.0	3815.0	4418.0	442.0	70.0	[69]
Elymus sibiricum	11.0	14.0	126.0	2374.0	185.0	50.0	0	[69]
Trifolium repens	26.5	35.6	45.7	415.3	167.7	1077.5	84.9	[64]
Trifolium striatum	–	–	37.0	214.4	443.5	1259.3	–	[58]
Trifolium subterraneum	–	23.7	240.5	503.9	369.9	2141.0	–	[53]
Ulex gallii	30.8	81.7	189.1	120.1	133.4	111.7	153.9	[64]
Calluna vulgaris	221.0	1190.0	474.0	203.0	450.0	829.0	–	[56]
Erica arborea	157.9	315.0	210.8	120.8	206.0	262.1	46.1	[64]
Erica tetralix	62.0	560.0	1124.0	1015.0	1496.0	4465.0	–	[56]
Vaccinium myrtillus	271.0	511.0	362.0	334.0	383.0	931.0	–	[56]

Table 3. Even-chain alcohol concentrations (mg/kg DM) of several herbaceous and shrub plant species.

4. Application of epicuticular compounds as biomarkers

The differences in the profiles of the epicuticular compounds mentioned above can be explored to estimate the proportions of different plant species and plant parts in different samples, such as herbage mixtures [70]; extrusa from oesophageal-fistulated animals [71] or faeces of sheep [52, 61, 65, 69, 72–75], goats [50, 60, 64, 76], cattle [62, 63, 66, 76, 77] and horses [62, 66, 77]. The principle of the application of the technique is simple and relies on the comparison of marker concentrations in a mixture (extrusa, digesta or faeces) and in diet components, plant species and/or plant parts that contribute (or could contribute) to that mixture. The comparison of the marker profiles can be made using different calculation procedures. It should be pointed out that more important than choosing the calculation procedure used, it is necessary to ensure that the information used (marker profiles of the possible diet components and the resultant mixture—faeces) is as accurate as possible.

Dove [70] proposed the utilization of simultaneous equations to estimate the proportions of the possible diet components when using alkanes as diet composition markers. In order to obtain unique solutions, the number of markers used is equal to the number of diet components and to the number of equations created [22]. The result of the equations indicates the amounts of the different diet components necessary to produce 1 kg of faeces, making possible to estimate the digestibility of the estimated diet. According to Dove and Mayes [78], this calculation procedure can be used in simple dietary mixtures, being more difficult to compute in complex mixtures. The main limitation of this procedure is in situations where there are more markers than the possible diet components, being necessary to select the markers to be used in the calculations. This selection involves arbitrary choices of the markers and the loss of information provided by the markers which were not used in the calculations. Moreover, this procedure may occasionally produce meaningless biological results as negative proportions of the diet components considered in the calculations.

In order to surpass these limitations, least-squares optimization methods can be applied, for which several algorithms have been developed [71, 79–81]. These calculation methodologies allow us to accommodate concentrations of different marker types (alkanes, LCFA, LCOH). The solution achieved by these algorithms attempts to minimize the squared deviations between the observed (O) marker concentrations in faeces and the concentration profile (E) arising from the diet composition estimate [21]:

$$\begin{aligned}\sum\nolimits_{i,j,k=1}^{n} [O-E]^2 = \sum\nolimits_{i,j,k=1}^{n} [&(Fi-(xAi+yBi+zCi+\dots)) \\ &+(Fj-(xAj+yBj+zCj+\dots)) \\ &+(Fk-(xAk+yBk+zCk+\dots))]^2 \text{ minimal}\end{aligned} \quad (1)$$

or

$$\begin{aligned}\sum\nolimits_{i,j,k=1}^{n} \left[O-E\right]^2 = \sum\nolimits_{i,j,k=1}^{n} \Bigg[&\left(\frac{Fi}{Ft}-\frac{xAi+yBi+zCi+\dots}{xAt+yBt+zCt+\dots}\right)+\left(\frac{Fj}{Fu}-\frac{xAj+yBj+zCj+\dots}{xAu+yBu+zCu+\dots}\right) \\ &+\left(\frac{Fk}{Fv}-\frac{xAk+yBk+zCk+\dots}{xAv+yBv+zCv+\dots}\right)\Bigg]^2 \text{ minimal}\end{aligned} \quad (2)$$

where x, y and z are the proportions of components A, B and C in the diet; Fi, Ai, Bi and Ci are the concentrations of alkane i in faeces and diet components A, B and C; Fj, Aj, Bj and Cj are the concentrations of LCOH j in faeces and diet components A, B and C; Fk, Ak, Bk and Ck are the concentrations of LCFA k in faeces and diet components A, B and C; Ft, At, Bt and Ct are total alkane concentrations in faeces and diet components A, B and C; Fu, Au, Bu and Cu are total LCOH concentrations in faeces and diet components *A*, *B* and *C*; *Fv*, *Av*, *Bv* and *Cv* are total LCFA concentrations in faeces and diet components *A*, *B* and *C*. It is possible to express the individual marker concentrations in the feeds and faeces as absolute concentrations (Eq. (1)) or as proportions of the total concentration (Eq. (2)). The advantage of using concentrations instead of proportions is that *x*, *y* and *z* estimates using Eq. (1) are the amounts which will result in 1 kg of faeces. Thus, this information can be used to obtain an estimate of diet digestibility as

$$\text{Dry matter digestibility} = \frac{(x+y+z+\ldots)-1}{(x+y+z+\ldots)} \quad (3)$$

5. Major constraints to the application of biomarkers in herbivory studies

As stressed by Dove and Mayes [21], it is important to ensure that the information used in both sides of Eqs. (1) and (2) (marker patterns of diet components and animal faeces) is as accurate as possible. An important source of error, often forgotten by researchers when applying the epicuticular markers to estimate diet selection in grazing studies, is the representativeness of the hand-collected samples of the vegetation components, in terms of marker profiles. This task can be difficult to accomplish as there can be significant variations in the marker profiles between plant species and plant parts within a specific plant species, as mentioned earlier. Other aspects requiring special attention are the continuous modification of each vegetation component available in the pasture, the relationship between plant parts and its stage of maturity and, consequently, their marker patterns. For this reason, it is recommended to collect samples of the plant species corresponding to each measuring period. Another important constraint associated with feeds/plant species is their very low marker concentrations. For example, herbaceous species (*L. perenne, T. repens, Pseudarrhenatherum longifolium, Agrostis capillaris,* [82] and *P. aquatica* [83]) and some shrub species (*U. gallii* [82]) are characterized by having low alkane concentrations and, for that reason, are more prone to analytical errors. Ferreira et al. [60] suggested that in these situations other marker types (e.g. LCFA or LCOH) should be used.

An additional concern is the collection of representative samples of faeces in terms of marker profiles. As occurs with other types of markers, the variation within and between days in the faecal concentrations can limit the utilization of this technique. In general, this variation is observed for dosed even-chain alkanes that are used for intake estimation [49, 84–91] due to their tendency to be associated with the liquid phase of the digesta. For that reason, an adaptation period of 5 days for the synthetic alkanes to reach a steady-state excretion pattern in animal faeces is generally suggested [21]. Regarding the natural markers, in grazing studies it is likely the existence of variation in the diet selected by the animals and, consequently, in feed intake, digestibility and faecal output from day to day [78]. For this reason, these authors suggest a sampling period of 5–7 days to obtain a more representative sample of faeces.

An assumption inherent to the application of the epicuticular compounds as diet composition markers is that they are totally recovered in the faeces. The results obtained in metabolic crate studies clearly indicate an incomplete recovery of alkanes [21, 73, 76], LCFA [59–63] and LCOH [52, 59, 64–66, 69] in the faeces of ruminant species (**Figure 3**), suggesting a close association between the length of the carbon chain of markers and its faecal recovery. Generally, results suggest a higher faecal recovery in the LCOH than in alkanes and LCFA [59, 64]. This is possibly related to the different location of these compounds in the wax layer (i.e. alkanes in greater concentrations in the epicuticular layer, whilst primary alcohols are found in greater concentrations

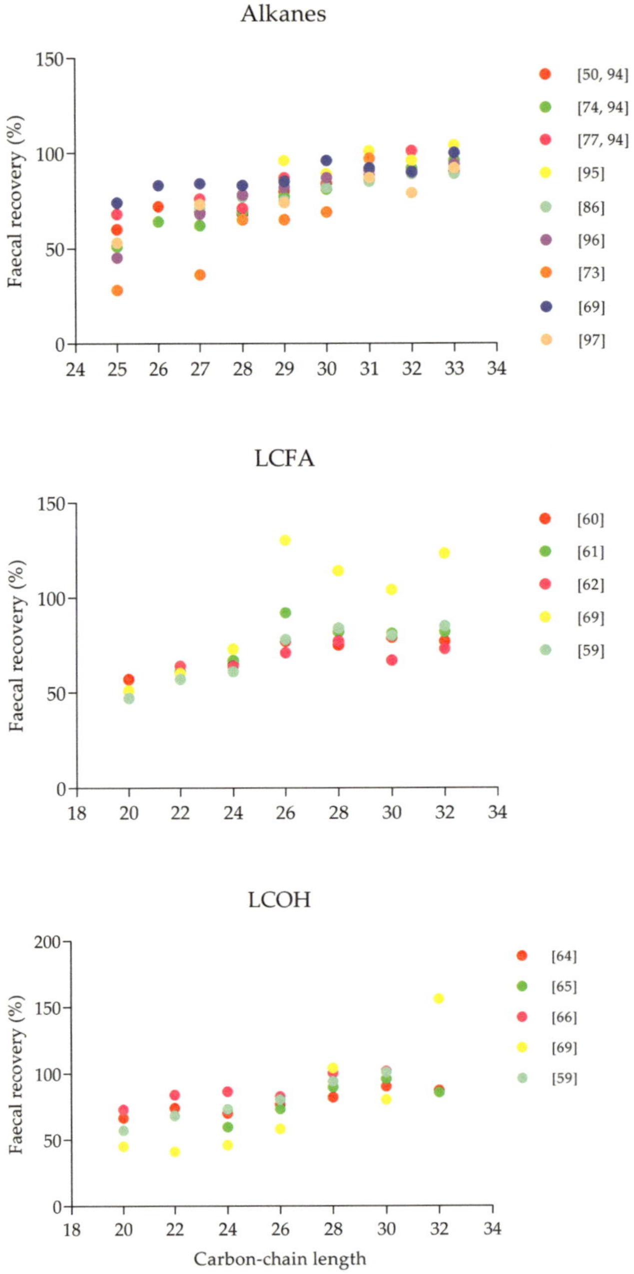

Figure 3. Effect of carbon-chain length on the faecal recovery (%) of alkanes [94], long-chain fatty acids (LFCA) and long-chain alcohols (LCOH) observed in ruminant species.

in the intracuticular wax [92] that could interfere with the efficiency of the extraction of these compounds and/or their absorption in the ruminants' digestive tract [65, 69, 93].

In several studies, the relationship between carbon-chain length and faecal recovery is better described by curvilinear functions for alkanes [60, 83, 93], LCFA [60, 61] and LCOH [65] as a result of the decrease in the difference of faecal recovery of markers with adjacent carbon-chain length with increasing carbon-chain length. By contrast, other studies have reported a linear association between carbon-chain length and faecal recovery in alkanes [75, 76, 96] and LCOH [64]. It seems that the association is dependent on the feeds/plant species comprising the diets and has an important effect on the accuracy of diet composition estimates on ruminant species. In fact, if uncorrected marker faecal concentrations are used in the calculations (Eq. (1) or (2)), estimates of diet composition will be biased towards those feeds/plant species with a predominance of longer carbon-chain length markers that have higher faecal recovery rates. For that reason, a suitable correction of marker faecal concentrations for incomplete faecal recovery, before applying them for diet composition estimation in grazing animals is generally suggested. Nevertheless, in situations where feeds/plant species do not have any chain-length bias, the effect of the recovery correction has little effect on the accuracy of diet composition estimates [60].

Data on marker faecal recoveries can be obtained in metabolic cage studies with animals fed on different mixtures of feeds/plant species that are available for each specific situation. It should be noted that in complex situations in terms of number of possible diet components, it will be difficult to decide which combination of plant species and/or plant parts will reflect the diet selected by each different animal species. For alkanes, Dove and Mayes [21] suggested another option that consists of dosing a range of synthetic even-chain alkanes and collecting the total faecal production by wearing faeces bags and calculating the faecal recoveries of natural odd-chain alkanes by interpolation. Nevertheless, it has been found that the synthetic dosed alkanes may have higher recoveries than those expected from interpolation of adjacent natural odd-chain alkanes [82, 87, 98].

For non-ruminant species such as horses [97, 99–102], pigs [103, 104], mountain hares [105] and pigeons [106], marker faecal recoveries seem to be unrelated to their carbon-chain length (**Figure 4**), indicating that these markers behave differently in the digestive tract of ruminants and non-ruminants, especially those with lower carbon-chain length. In fact, the comparison of faecal recovery data between ruminant and non-ruminant species for alkanes [77], LCFA [62] and LCOH [66] indicates a greater disappearance of the shorter markers in the gut of ruminants than in non-ruminants. The site and the mechanisms underlying marker losses in the animal gastrointestinal tract are still far from being completely elucidated. Earlier studies undertaken by Mayes et al. [107] suggested that the disappearance of the dosed alkanes C_{28}, C_{32} and C_{36} occurred mainly in the small intestine in sheep. More recently, Keli et al. [25] also suggested that alkane disappearance should mainly occur in the small intestine as they were not be able to find evidences of rumen microorganisms' capability to synthesize or metabolize alkanes in in vitro conditions. By contrast, Ohajuruka and Palmquist [108] found that the loss of dosed C_{32} alkane in dairy cows occurs mainly in the rumen.

The lack of a clear relationship between the carbon-chain length of the epicuticular compounds and their faecal recovery in non-ruminant species has an important effect on diet composition estimates. In fact, Ferreira et al. [77] were not be able to observe an increase in the accuracy of diet composition estimates in horses when alkane faecal concentrations corrected for their incomplete recovery were used. This lower dependence of markers for a suitable faecal recovery correction was also found in LCOH and LCFA by López López et al. [66] and Ferreira et al. [62], respectively, although in their cases a linear association between carbon-chain length and faecal recovery was observed. These results indicate that, for this animal species, accurate estimates of diet composition can be obtained even when raw data of the faecal

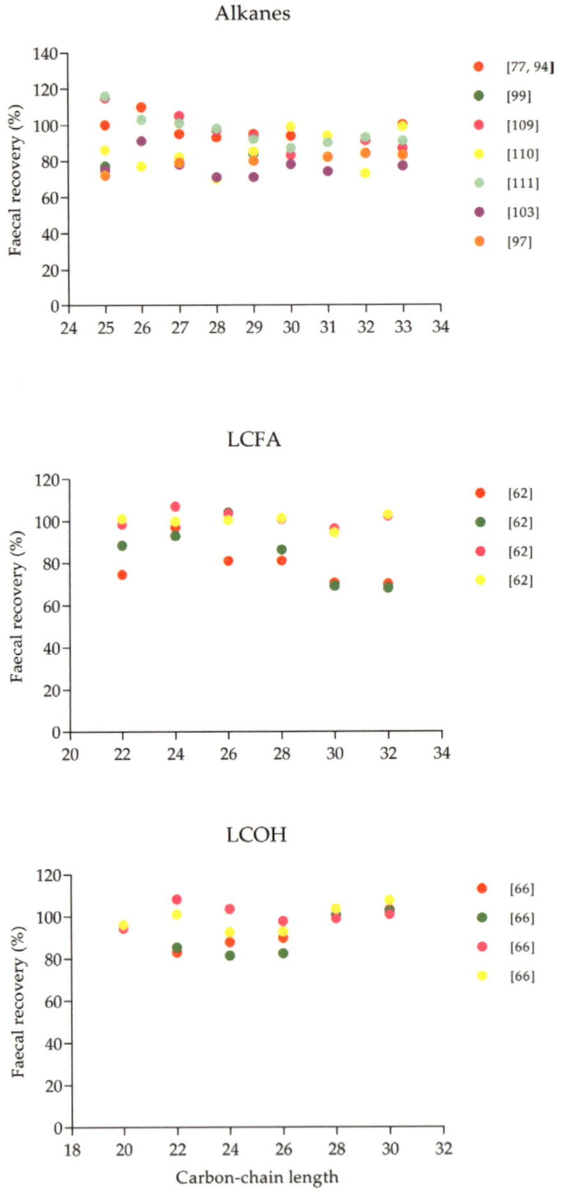

Figure 4. Effect of carbon-chain length on the faecal recovery (%) of alkanes [109–111], long-chain fatty acids (LFCA) and long-chain alcohols (LCOH) observed in non-ruminant species.

concentrations of these epicuticular compounds (i.e. without previous corrections of the faecal concentrations) are used.

Another important constraint that limits a wider applicability of epicuticular compounds as markers in grazing studies is that their faecal recovery may depend on the diet composition, compelling researchers to calculate faecal recoveries for each specific situation (i.e. diet composition), making it impossible to use recovery data available in literature. This effect was observed in several studies performed with alkanes [50, 83], LCOH [64, 65, 69] and LCFA [60, 61, 69], whilst others were not be able to detect it [52, 73, 76, 96]. According to some authors [64, 69, 96], this inconsistency may be due to the particular plant species comprising the diets. Lin et al. [69, 75], using sheep fed distinct grass species (*E. sibiricum*, *L. chinensis* or *L. dasystachys*) obtained different faecal recoveries of LCOH, LCFA and alkanes. According to the same authors, these results could be explained by differences among plant species in their plant cuticular wax morphology, influencing the level of extraction and the absorption of these compounds in the digestive tract of animals. Diet digestibility may also explain these differences in the faecal recovery of the epicuticular compounds among different diets. In fact, a general tendency for higher faecal recoveries of alkanes [50, 74, 112] and LCOH [64] in diets with lower digestibility was observed. Lower accessibility to the cuticular waxes of those feeds as a result of a higher association of cuticle with cell wall components [113] may explain the lower availability of epicuticular compounds to be absorbed in the digestive tract of animals.

The first epicuticular compounds suggested as diet composition markers were the alkanes [45], and the limited number of components (e.g. plant species and/or plant parts) that can be discriminated in the diet that is restricted to the number of n-alkanes available was soon recognized. The number of alkanes available for diet composition calculations is generally limited to 9 (C_{25}–C_{33}), due to the higher potential analytical error associated with those of very low concentrations both in plants and in faeces, which may contribute to the discrepancy in sum of squares in the calculation method [78, 95]. Moreover, it is accepted that the increase of the number of diet components to be discriminated will likely result in less accurate diet composition estimates, as it increases the likelihood that an observed alkane pattern in faeces may result from different combinations of diet components [21]. To overcome these limitations, a possible approach to obtain reliable diet composition estimates is to increase the number of 'discriminators' by combining the use of alkanes with other plant-wax markers, such as alkenes [67], LCOH [52, 53, 64, 69] and LCFA [59–63]. According to Bugalho et al. [53], combination of markers should only be performed when additional discriminatory information is provided. Combination of different marker types may improve the accuracy of diet composition estimates as markers with greater concentrations, less prone to analytical error, can be selected in situations where the possible dietary feed items have similar alkane profiles. Moreover, it is likely that the use of a greater number of markers provides a more specific 'fingerprint' for a particular plant component [60]. This is also very important even when the number of plant species to be discriminated is low, but they present high similarities in alkane profiles, making it difficult to discriminate them.

In several studies, it was possible to observe an increase in the accuracy of diet composition estimates when combining two [52, 60, 114] or three marker types [64–66, 68, 69]. It should be pointed out that the combination of epicuticular compounds does not necessarily result in more accurate estimates of diet composition [21, 53]. For example, Vargas-Jurado et al. [115] did not observe an improvement of the predictions of the composition of mixtures of tall fescue (*Festuca arundinacea*) and red clover (*Trifolium pratense*) when LCOH was combined with alkanes. Nevertheless, results obtained by Ferreira et al. [60, 64, 65] suggest that the combination of markers reduce their dependence on accurate faecal-correction data for an accurate diet composition estimation as more specific fingerprints of each plant species are achieved, increasing their ability to discriminate them.

Another approach that is suggested when the number of possible diet components is high is to decrease the number of possible diet components by pooling the available plant species into groups [81, 116]. These groups are formed by plant species with similar marker profiles, based on multivariate statistical analysis, that are then treated as dietary components in the calculations. One aspect that needs particular attention is the fact that the accuracy of diet composition estimates can be influenced by different availability or selectivity levels of some plant species within each group, especially if the marker profile of a particular plant species is distinct from the mean marker profile of the group in which that species is included [51]. As pointed out by Ferreira et al. [116], feeding selectivity effect will depend on the particular species that could be selected within the group and on the similarity in the marker profile of the plant species of the dietary group. Bugalho et al. [51] did not found any feeding selectivity effect within a group of 19 herbaceous species on diet composition estimates of red deer. Similar results were observed by Ferreira et al. [116] when applying different levels of feeding selectivity to a dietary group formed by heather species (*C. vulgaris*, *Erica cinerea*, *Erica umbellata* and *Erica australis*) with similar alkane profile. By contrast, when the same procedure was applied to a dietary group formed by three grass species (*L. perenne*, *P. longifolium* and *A. capillaris*), significant modifications of the proportions of each dietary group were observed. A decrease in feeding selectivity effect can be achieved by the formation of more uniform dietary groups in terms of marker profiles. However, Ferreira et al. [63] observed higher levels of accuracy of diet composition estimates when considering all plant species that animals had at their disposal compared to its grouping according to the similarity of alkane and LCFA profiles.

The exclusion of plant species based on preliminary information is another approach suggested by Dove and Mayes [21] to reduce the number of possible diet components. The observation of the animals' feeding behaviour, plant-derived data or information based on other methodologies that indicate the rejection of a particular plant species and/or vegetation community, can help the researcher to use more accurate data on the plant species that should be considered in the calculations.

The utilization/combination of different types of markers has also advantages in less complex plant communities (i.e. lower number of plant species to be discriminated), by giving the opportunity to the researcher to choose those with higher concentrations less prone to measurement errors in their analytical determination [21]. According to Charmley and Dove [83] and Ferreira et al. [60, 61], the utilization of markers with low concentrations can turn discrimination of plant species more difficult and may result in less accurate estimates of diet composition. In fact, Oliván et al. [82] attributed the difficulties in distinguishing three grass species (*L. perenne*, *P. longifolium* and *A. capillaris*) and gorse to their low alkane concentrations. Also, Charmley and Dove [83] had difficulties in obtaining accurate estimates of diet composition when *P. aquatica* (plant species characterized by very low alkane concentrations) was a component of diets fed to sheep. In our opinion, the exclusion of markers based on their low concentrations should be performed with caution as, in some situations, they may dis-

http://dx.doi.org/10.5772/67345

criminate better plant species than those with higher concentrations. Thus, Dove and Mayes [21] suggested that the balance between the capability of markers to discriminate plant species and the level of potential analytical error should be considered when choosing markers.

6. Conclusions

Taking into account all the data presented in this chapter, it is certain that the application of the epicuticular compounds as faecal markers can improve our knowledge on the grazing behaviour, particularly diet selection, of free-ranging herbivore species under different vegetation conditions. Although some shortcomings can be pointed out to these faecal markers, namely the variation of profiles within plant species and morphological parts, lack of inertness in the digestive tract of ruminant species and, for that reason, the need for a suitable recovery correction of their faecal concentrations, they have been used quite successfully. Its application allows to overcome major limitations recognized to the traditional techniques in terms of accuracy and extent of the results (i.e. identification of plant species and/or plant parts), animal welfare issues (i.e. avoid the need for fistulated animals; lower disturbance of animals compromising its normal grazing behaviour) and intensive labour. The combination of different maker types (alkanes, LCOH, LCFA) seems promising to overcome the enumerated constraints and to extend their application to more complex vegetation communities. Therefore, research on the identification of other chemical compounds should continue to be developed. Finally, data obtained from different available techniques (microhistological procedures, NIRS, Fourier-transform infrared (FTIR) spectroscopy and fluorescence spectroscopy, telemetry solutions) should be integrated in order to enhance the accuracy of diet composition being selected by herbivore species. This will further improve the precision of information (i.e. possible diet components) used when applying the epicuticular markers.

Acknowledgements

This work was financed by the Portuguese Science and Technology Foundation (FCT) under the Project UID/CVT/00772/2013.

Author details

Luis Miguel Mendes Ferreira[1*], Rafael Celaya[2], Ana Sofia Santos[3], Koldo Osoro[2] and Miguel Rodrigues[1]

*Address all correspondence to: lmf@utad.pt

1 CECAV—Animal and Veterinary Research Centre, Universidade de Trás-os-Montes e Alto Douro (UTAD), Quinta de Prados, Vila Real, Portugal

2 Regional Service of Agrifood Research and Development (SERIDA), Villaviciosa, Asturias, Spain

3 CITAB—Centre for the Research and Technology of Agro-Environmental and Biological Sciences, University of Trás-os-Montes and Alto Douro, Quinta de Prados, Vila Real, Portugal

References

[1] Gordon IJ, Hester AJ, Festa-Bianchet M. The management of wild large herbivores to meet economic, conservation and environmental objectives. J. Appl. Ecol. 2004;41:1021–1031. DOI:10.1111/j.0021-8901.2004.00985.x

[2] Hodgson J. Grazing behaviour and herbage intake. In: Frame J, editor. Grazing. BGS Occ. Symp. No. 19;1986. pp. 51–64.

[3] Hutchings NJ, Gordon IJ. A dynamic model of herbivore-plant interactions on grasslands. Ecol. Model. 2001;136:209–222. DOI:10.1016/S0304-3800(00)00426-9

[4] Hester AJ, Gordon IJ, Baillie GJ, Tappin E. Foraging behaviour of sheep and red deer within natural heather/grass mosaics. J. Appl. Ecol. 1999;36:133–146. DOI:10.1046/j.1365-2664.1999.00387.x

[5] Hülber K, Ertl S, Gottfried M, Reiter K, Grabherr G. Gourmets or gourmands? Diet selection by large ungulates in high-alpine plant communities and possible impacts on plant propagation. Basic Appl. Ecol. 2005;6:1–10. DOI:10.1016/j.baae.2004.09.010

[6] Rook AJ, Dumont B, Isselstein J, Osoro K, WallisDeVries MF, Parente G, Mills J. Matching type of grazing animal to desired biodiversity outcomes—a review. Biol. Conserv. 2004;119:137–150. DOI:10.1016/j.biocon.2003.11.010

[7] Rook AJ, Tallowin JRB. Grazing and pasture management for biodiversity benefit. Anim. Res. 2003;52:181–189. DOI:10.1051/animres:2003014

[8] Hodgson J, Forbes TDA, Armstrong RH, Beattie MM, Hunter EA. Comparative studies of the ingestive behaviour and herbage intake of sheep and cattle grazing indigenous hill plant communities. J. Appl. Ecol. 1991;28:205–227. DOI:10.2307/2404126

[9] Celaya R, Oliván M, Ferreira LMM, Martínez A, García U, Osoro K. Comparison of grazing behaviour, dietary overlap and performance in non-lactating domestic ruminants grazing on marginal heathland areas. Livest. Sci. 2007;106:271–281. DOI:10.1016/j.livsci.2006.08.013

[10] Fraser MD, Theobald VJ, Griffiths JB, Morris SM, Moorby JM. Comparative diet selection by cattle and sheep grazing two contrasting heathland communities. Agric. Ecosyst. Environ. 2009;129:182–192. DOI:10.1016/j.agee.2008.08.013

[11] Osoro K, Oliván M, Celaya R, Martínez A. Effects of genotype on the performance and intake characteristics of sheep grazing contrasting hill vegetation communities. Anim. Sci. 1999;69:419–426.

[12] Osoro K, García U, Jáuregui BM, Ferreira LMM, Rook AJ, Celaya R. Diet selection and live-weight changes of two breeds of goats grazing on heathlands. Animal 2007;1:449–457. DOI:10.1017/S1751731107683797

[13] Hofmann RR. Evolutionary steps of ecophysiological adaptation and diversification of ruminants: comparative view of their digestive system. Oecologia 1989;78:443–457. DOI:10.1007/BF00378733

[14] Mysterud A. The relative roles of body size and feeding type on activity time of temperate ruminants. Oecologia 1998;113:442–446. DOI:10.1007/s004420050396

[15] Pérez-Barbería FJ, Gordon IJ, Illius AW. Phylogenetic analysis of stomach adaptation in digestive strategies in African ruminants. Oecologia 2001;129:498–508. DOI:10.1007/s004420100768

[16] Van Soest PJ. Nutritional Ecology of the Ruminant. 2nd ed. Ithaca, NY: Cornell University Press; 1994. 476p.

[17] Udén P, Van Soest PJ. Comparative digestion of timothy (*Phleum pratense*) fibre by ruminants, equines and rabbits. Br. J. Nutr. 1982;47:267–272. DOI:10.1079/BJN19820035

[18] Ferreira LMM, Celaya R, Benavides R, Jáuregui BM, García U, Santos AS, Rosa García R, Rodrigues MAM, Osoro K. Foraging behaviour of domestic herbivore species grazing on heathlands associated with improved pasture areas. Livest. Sci. 2013;155:373–383. DOI:10.1016/j.livsci.2013.05.007

[19] Osoro K, Ferreira LMM, García U, Martínez A, Celaya R. Forage intake, digestibility and performance of cattle, horses, sheep and goats grazing together on an improved heathland Anim. Prod. Sci. DOI:10.1071/AN15153

[20] Holechek JL, Vavra M, Pieper RD. Botanical composition determination of range herbivore diets: a review. J. Range Manage. 1982;35:309–315. DOI:10.2307/3898308

[21] Dove H, Mayes RW. Using n-alkanes and other plant wax components to estimate intake, digestibility and diet composition of grazing/browsing sheep and goats. Small Rumin. Res. 2005;59:123–139. DOI:10.1016/j.smallrumres.2005.05.016

[22] Dove H, Mayes RW. Plant wax components: A new approach to estimating intake and diet composition in herbivores. J. Nutr. 1996;126:13–26.

[23] Mayes RW, Dove H. Measurement of dietary nutrient intake in free-ranging mammalian herbivores. Nutr. Res. Rev. 2000;13:107–138. DOI:10.1079/095442200108729025

[24] Walker JW, Campbell ES, Lupton CJ, Taylor CA Jr, Waldron DF, Landau S. Effects of breed, sex, and age on the variation and ability of Fecal near-infrared reflectance spectra to predict the composition of goat diets. J. Anim. Sci. 2007;85:518–526. DOI:10.2527/jas.2006-202

[25] Keli A, Mayes RW, de Vega A. *In vitro* studies of the metabolism of [^{14}C]-n-alkanes using ruminal fluid of sheep as substrate. Animal 2008;2:1748–1752. DOI:10.1017/S1751731108003108

[26] Decruyenaere V, Buldgen A, Stilmant D. Factors affecting intake by grazing ruminants and related quantification methods: a review. Biotechnol. Agron. Soc. Environ. 2009;13:559–573.

[27] Landau SY, Dvash L, Roudman M, Muklada H, Barkai D, Yehuda Y, Ungar ED. Faecal near-IR spectroscopy to determine the nutritional value of diets consumed by beef cattle in east Mediterranean rangelands. Animal 2016;10:192–202. DOI:10.1017/S175173111500169X

[28] Dixon RM, Coates DB. Review: Near infrared spectroscopy of faeces to evaluate the nutrition and physiology of herbivores. J. Near Infrared Spectrosc. 2009;17:1–31. DOI:10.1255/jnirs.822

[29] Swain DL, Friend MA. Opportunities for telemetry techniques in studies on the nutritional ecology of free-ranging domesticated ruminants. Animal 2013;7(s1):123–131. DOI:10.1017/S1751731112000870

[30] Vallentine JF. Grazing Management. 2nd ed. San Diego, CA: Academic Press; 2001. 659 p.

[31] Hebblewhite M, Haydon DT. Distinguishing technology from biology: a critical review of the use of GPS telemetry data in ecology. Philos. Trans. R. Soc. Lond. B Biol. Sci. 2010;365:2303–2312. DOI: 10.1098/rstb.2010.0087

[32] Swain DL, Wark T, Bishop-Hurley GJ. Using high fix rate GPS data to determine the relationships between fix rate, prediction errors and patch selection. Ecol. Model. 2008;212:273–279. DOI: 10.1016/j.ecolmodel.2007.10.027

[33] Thompson DJ, Wheatley BJ, Church JS, Newman R, Walker J. Comparing grazing and resting electivity of beef cattle for BC bunchgrass communities using GPS collars. Can. J. Anim. Sci. 2015;95:499–507. DOI:10.4141/cjas-2014-116

[34] Jones RJ, Ludlow MM, Troughton JH, Blunt CG. Estimation of the proportion of C_3 and C_4 plant species in the diet of animals from the ratio of natural ^{12}C and ^{13}C isotopes in the faeces. J. Agric. Sci. 1979;92:91–100. DOI: 10.1017/S0021859600060536

[35] García SC, Holmes CW, Hodgson J, MacDonald A. The combination of the n-alkanes and ^{13}C techniques to estimate individual dry matter intakes of herbage and maize silage by grazing dairy cows. J. Agric. Sci. 2000;135:47–55.

[36] Sponheimer M, Robinson T, Ayliffe L, Passey B, Roeder B, Shipley L, Lopez E, Cerling T, Dearing D, Ehleringer J. An experimental study of carbon-isotope fractionation between diet, hair, and feces of mammalian herbivores. Can. J. Zool. 2003;81:871–876. DOI: 10.1139/Z03-066

[37] Norman HC, Wilmot MG, Thomas DT, Masters DG, Revell DK. Stable carbon isotopes accurately predict diet selection by sheep fed mixtures of C_3 annual pastures and saltbush or C_4 perennial grasses. Livest. Sci. 2009;121:162–172. DOI: 10.1016/j.livsci.2008.06.005

[38] Botha S, Stock W. Stable isotope composition of faeces as an indicator of seasonal diet selection in wild herbivores in southern Africa. S. Afr. J. Sci. 2005;101:371–374

[39] De Smet S, Balcaen A, Claeys E, Boeckx P, van Cleemput O. Stable carbon isotope analysis of different tissues of beef animals in relation to their diet. Rapid Commun. Mass Spectrom. 2004;18:1227–1232. DOI: 10.1002/rcm.1471

[40] Post-Beittenmiller D. Biochemistry and molecular biology of wax production in plants. Annu. Rev. Plant Physiol. Plant Mol. Biol. 1996;47:405–430. DOI:10.1146/annurev.arplant.47.1.405

[41] Dove H, Mayes RW. Protocol for the analysis of n-alkanes and other plant-wax compounds and for their use as markers for quantifying the nutrient supply of large mammalian herbivores. Nat. Protoc. 2006;4:1680–1697. DOI:10.1038/nprot.2006.225

[42] Eigenbrode SD, Espelie KE. Effects of plant epicuticular lipids on insect herbivores. Annu. Rev. Entomol. 1995;40:171–194. DOI:10.1146/annurev.en.40.010195.001131

[43] Body DR, Hansen RP. The occurrence of C13 to C31 branched-chain fatty acids in the faeces of sheep fed rye grass, and of C12 to C34 normal acids in both the faeces and the rye grass. J. Sci. Food Agric. 1978;29:107–14. DOI:10.1002/jsfa.2740290206

[44] Grace ND, Body DR. The possible use of long chain (C_{19}-C_{32}) fatty acids in herbage as an indigestible faecal marker. J. Agric. Sci. 1981;97:743–745. DOI: 10.1017/S0021859600037126

[45] Dove H, Mayes RW. The use of plant wax alkanes as markers substances in studies of the nutrition of herbivores: a review. Aust. J. Agric. Res. 1991;42:913–957. DOI:10.1071/AR9910913

[46] Baker SK, Klein L. Potential to use n-alkanes in plant cuticular waxes to discriminate plant parts of subterranean clovers eaten by ruminants. Proc. Aust. Soc. Anim. Prod. 1994;20:419. Available from: http://www.asap.asn.au/livestocklibrary/1994/Baker94.PDF [Accessed: 2016-07-12]

[47] Dove H, Mayes RW, Freer M. Effects of species, plant part, and plant age on the n-alkane concentrations in the cuticular wax of pasture plants. Aust. J. Agric. Res. 1996;47:1333–47. DOI:10.1071/AR9961333

[48] Laredo MA, Simpson GD, Minson DJ, Orpin CG. The potential for using n-alkanes in tropical forages as a marker for determination of dry matter intake by grazing ruminants. J. Agric. Sci. 1991;117:355–361. DOI: 10.1017/S0021859600067101

[49] Morais JAS, Berchielli TT, de Veja A, Queiroz MFS, Keli A, Reis RA, Bertipaglia LMA, Souza SF. The validity of *n*-alkanes to estimate intake and digestibility in Nellore beef cattle fed a tropical grass (*Brachiaria brizantha* cv. Marandu). Livest. Sci. 2011;135:184–192. DOI: 10.1016/j.livsci.2010.07.004

[50] Ferreira LMM, Oliván M, Rodrigues MAM, García U, Osoro K. Validation of the alkane technique to estimate diet selection of goats grazing heather-gorse vegetation communities. J. Sci. Food Agric. 2005;85:1636–1646. DOI:10.1002/jsfa.2162

[51] Bugalho M, Mayes RW, Milne JA. The effects of feeding selectivity on the estimation of diet composition using the n-alkane technique. Grass Forage Sci. 2002;57:224–231. DOI: 10.1046/j.1365-2494.2002.00320.x

[52] Dove H, Charmley E. Using the alkanes and long-chain alcohols of plant cuticular wax to estimate diet composition and the intakes of mixed forages in sheep consuming a known amount of alkane-labelled supplement. Animal. 2008;2:1474–1485. DOI:10.1017/S1751731108002735

[53] Bugalho MN, Dove H, Kelman W, Wood JT, Mayes RW. Plant wax alkanes and alcohols as herbivore diet composition markers. J. Range Manage. 2004;57:259–268. DOI:10.2458/azu_jrm_v57i3_bugalho

[54] Oliván M, Osoro K. Utilization of the n-alkane technique to study intake and diet selection of grazing ruminants: review. ITEA Inf. Tec. Econ. Agrar. 1997;93A:193–208.

[55] Lin LJ, Luo HL, Wang H, Zhang YJ, Shu B. Evaluation of long-chain alcohols and fatty acids, in combination with alkanes, as markers in the estimation of the composition of four herbages in mixtures. Grass Forage Sci. 2008;64;19–25. DOI:10.1111/j.1365-2494.2008.00663.x

[56] Fraser MD, Theoblad VJ, Moorby JM. Determining diet composition on complex swards using n-alkanes and long-chain fatty alcohols. Ecol. Appl. 2006;16:1901–1910. DOI: 10.1890/1051-0761(2006)016[1901:DDCOCS]2.0.CO;2

[57] Oliveira DE, Prates ER, Peralba MCR. Identification and quantification of n-alkanes of forage plant waxes. Rev. Bras. Zootec. 1997;26:881–886.

[58] Smith DG, Mayes RW, Raats JG. Effects of species, plant part, and season of harvest on n-alkane concentrations in the cuticular wax of common rangeland grasses from southern Africa. Aust. J. Agric. Res. 2001;52:875–882. DOI:10.1071/AR00032

[59] Ali HAM, Mayes RW, Lamb CS, Hector BL, Verma AK, Orskov ER. The potential of long-chain fatty alcohols and long-chain fatty acids as diet composition markers: development of methods for quantitative analysis and faecal recoveries of these compounds in sheep fed mixed diets. J. Agric. Sci. 2004;142:71–78. DOI:10.1017/S0021859604004034

[60] Ferreira LMM, Carvalho S, Falco V, Celaya R, García U, Santos AS, Rodrigues MAM, Osoro K. Assessment of very long-chain fatty acids as complementary or alternative natural fecal markers to n-alkanes for estimating diet composition of goats feeding on mixed diets. J. Anim. Sci. 2009;87:2732–2745. DOI:10.2527/jas.2008-1718

[61] Ferreira LMM, Celaya R, Falco V, Oliván M, Santos AS, Guedes C, Rodrigues MAM, Osoro K. Evaluation of very long-chain fatty acids and n-alkane epicuticular compounds as markers for estimating diet composition of sheep fed heathland vegetation species. Anim. Feed Sci. Technol. 2010;156:75–88. DOI: 10.1016/j.anifeedsci.2010.01.007

[62] Ferreira LMM, Celaya R, Santos AS, Falco V, Guedes C, Rodrigues MAM, Osoro K. Comparison of long-chain fatty acids and alkanes as markers to estimate diet composition of equines and cattle consuming heathland vegetation species. Livest. Sci. 2010;131:260–271. DOI:10.1016/j.livsci.2010.04.011

[63] Ferreira LMM, Celaya R, Santos AS, Falco V, Guedes C, Rodrigues MAM, Osoro K. The utilization of long-chain fatty acids as markers for diet composition estimates in ruminants: effects of animal species, diet composition and marker combination. Grass Forage Sci. 2011;66:183–195. DOI:10.1111/j.1365-2494.2010.00774.x

[64] Ferreira LMM, Celaya R, Santos AS, Guedes CMV, Rodrigues MAM, Mayes RW, Osoro K. Evaluation of long-chain alcohols as diet composition markers in goats grazing heathland areas. Animal 2012;6:683–692. DOI:10.1017/S1751731111001881

[65] Ferreira LMM, Celaya R, Santos AS, Mayes RW, Rodrigues MAM, Osoro K. Application of long-chain alcohols as diet composition markers in sheep fed on grass–clover/heather–gorse plant species. Grass Forage Sci. 2015;70:30–43. DOI:10.1111/gfs.12083

[66] López López CL, Celaya R, Santos AS, Rodrigues MAM, Osoro K, Ferreira LMM. Application of long-chain alcohols as faecal markers to estimate diet composition of equines and cattle fed with herbaceous and woody species. Animal 2015;9:1786–1794. DOI: 10.1017/S1751731115001196

[67] Dove H, Oliván M. The possible use of the alkenes (unsaturated hydrocarbons) of plant cuticular wax as diet composition markers in sheep. In: Sandoval-Castro CA, DeB Hovell FD, Torres-Acosta JFJ, Ayala-Burgos A, editors. Herbivores. The Assessment of Intake, Digestibility and the Roles of Secondary Compounds. England: Nottingham University Press; 2006. pp 1–9.

[68] Ali HAM, Mayes RW, Hector BL, Ørskov ER. Assessment of n-alkanes, long-chain fatty alcohols and long-chain fatty acids as diet composition markers: the concentrations of these compounds in rangeland species from Sudan. Anim. Feed Sci. Technol. 2005;121:257–271. DOI: 10.1016/j.anifeedsci.2005.02.026

[69] Lin LJ, Luo HL, Zhang YJ, Wang H, Shu B, Hong FZ. The potential use of long-chain alcohols and fatty acids as diet composition markers: factors influencing faecal recovery rates and diet composition estimates in sheep. Animal. 2009;3:1605–1612. DOI:10.1017/S1751731109990401

[70] Dove H. Using the n-alkanes of plant cuticular wax to estimate the species composition of herbage mixtures. Aust. J. Agric. Res. 1992;43:1711–1724. DOI:10.1071/AR9921711

[71] Salt CA, Mayes RW, Colgrove PM, Lamb CS. The effects of season and diet composition on the radiocaesium intake by sheep grazing on heather moorland. J. Appl. Ecol. 1994;31:125–136. DOI:10.2307/2404605

[72] Lewis RM, Magadlela AM, Jessop NS, Emmans GC. The ability of the n-alkane technique to estimate intake and diet choice of sheep. Anim. Sci. 2003;77:319–327.

[73] Valiente OL, Delgado P, de Vega A, Guada JA. Validation of the n-alkane technique to estimate intake, digestibility, and diet composition in sheep consuming mixed grain: roughage diets. Aust. J. Agric. Res. 2003;54:693–702. DOI:10.1071/AR02221

[74] Ferreira LMM, Oliván M, Celaya R, Garcia U, Rodrigues MAM, Osoro K. The use of the alkane technique to estimate diet selection of sheep grazing grass-clover/heather-gorse vegetation communities. J. Sci. Food Agric. 2007;87:274–285. DOI:10.1002/jsfa.2717

[75] Lin LJ, Luo HL, Zhang YJ, Shu B. The effects, in sheep, of dietary plant species and animal live weight on the faecal recovery rates of alkanes and the accuracy of intake and diet composition estimates obtained using alkanes as faecal markers. J. Agric. Sci. 2007;145:87–94. DOI:10.1017/S002185960600654X

[76] Brosh A, Henkin Z, Rothman SJ, Aharoni Y, Orlov A, Arieli A. Effects of faecal n-alkane recovery in estimates of diet composition. J. Agric. Sci. 2003;140:93–100. DOI:10.1017/S0021859602002757

[77] Ferreira LMM, Garcia U, Rodrigues MAM, Celaya R, Dias-da-Silva A, Osoro K. The application of the *n*-alkane technique for estimating the composition of diets consumed by equines and cattle feeding on upland vegetation communities. Anim. Feed Sci. Technol. 2007;138:47–60. DOI: 10.1016/j.anifeedsci.2006.11.007

[78] Dove H, Mayes RW. Wild and Domestic Herbivore Diet Characterization. Satellite Meeting of the VI International Symposium on the Nutrition of Herbivores; 19–24 October 2003; Mérida, Yucatán, México. 2003. 88 p.

[79] Dove H, Moore AD. Using a least-squares optimization procedure to estimate botanical composition based on the alkanes of plant cuticular wax. Aust. J. Agric. Res. 1995;46:1535–1544. DOI:10.1071/AR9951535

[80] Newman JA, Thompson WA, Penning PD, Mayes RW. Least-squares estimation of diet composition from n-alkanes in herbage and faeces using matrix mathematics. Aust. J. Agric. Res. 1995;46:793–805. DOI:10.1071/AR9950793

[81] Martins H, Elston DA, Mayes RW, Milne JA. Assessment of the use of n-alkanes as markers to describe the complex diets of herbivores. J. Agric. Sci. 2002;138;425–434. DOI:10.1017/S0021859602002046

[82] Oliván M, Ferreira LMM, Celaya R, Osoro K. Accuracy of the n-alkane technique for intake estimates in beef cattle using different sampling times and feeding levels. Livest. Sci. 2007;106:28–40. DOI:10.1016/j.livsci.2006.06.015

[83] Charmley E, Dove H. Using plant wax markers to estimate diet composition and intakes of mixed forages in sheep by feeding a known amount of alkane-labelled supplement. Aust. J. Agric. Res. 2007;58:1215–1225. DOI: 10.1071/AR07187

[84] Azevedo EB, Poli CHEC, David DB, Amaral GA, Fonseca L, Carvalho PCF, Fischer V, Morris ST. Use of faecal components as markers to estimate intake and digestibility of grazing sheep. Livest. Sci. 2014;165:42–50. DOI:10.1016/j.livsci.2014.04.018

[85] Bezabih M, Pellikaan WF, Tolera A, Hendriks WH. Estimation of feed intake and digestibility in cattle consuming low-quality tropical roughage diets using molasses-based n-alkane boluses. Anim. Feed Sci. Technol. 2012;177:161–171. DOI: 10.1016/j.anifeedsci.2012.08.014

[86] Mayes RW, Lamb CS, Colgrove PM. The use of dosed and herbage n-alkanes as markers for the determination of herbage intake. J. Agric. Sci. 1986;107:161–170. DOI:10.1017/S0021859600066910

[87] Hendricksen RE, Reich MM, Roberton RF, Reid DJ, Gazzola C, Rideout JA, Hill RA. Estimating the voluntary intake and digestibility of buffel-grass and lucerne hays offered to Brahman-cross cattle using n-alkanes. Anim. Sci. 2002;74:567–577.

[88] Elwert C, Dove H. Estimation of roughage intake in sheep using a known daily intake of a labelled supplement. Anim. Sci. 2005;81:47–56. DOI:10.1079/ASC41940047

[89] Narvaez N, Brosh A, Pittroff W. Use of n-alkanes to estimate seasonal diet composition and intake of sheep and goats grazing in California chaparral. Small Rumin. Res. 2012;104:129–138. DOI:10.1016/j.smallrumres.2011.10.002

[90] Ferreira LMM, Oliván M, Rodrigues MAM, Osoro K, Dove H, Dias-da-Silva A. Estimation of feed intake by cattle using controlled-release capsules containing n-alkanes or chromium sesquioxide. J. Agric. Sci. 2004;142:225–234. DOI:10.1017/S0021859604004320

[91] Ferreira LMM, Garcia U, Rodrigues MAM, Celaya R, Dias-da-Silva A, Osoro K. Estimation of feed intake and apparent digestibility of equines and cattle grazing on

heathland vegetation communities using the n-alkane markers. Livest. Sci. 2007;110:46–56. DOI:10.1016/j.livsci.2006.09.026

[92] Buschhaus C, Herz H, Jette R. Chemical composition of the epicuticular and intracuticular wax layers on adaxial sides of *Rosa canina* leaves. Ann. Bot. (Lond.) 2007;100:1557–1564. DOI: 10.1093/aob/mcm255

[93] Elwert C, Dove H, Rodehutscord M. Faecal alkane recoveries from multi-component diets and effects on estimates of diet composition in sheep. Animal 2008;2:125–134. DOI:10.1017/S1751731107000900

[94] Ferreira LMM, Celaya R, García U, Rodrigues MAM, Osoro K. Differences between domestic herbivores species in alkane faecal recoveries and the accuracy of subsequent estimates of diet composition. Anim. Feed Sci. Technol. 2009;151:128–142. DOI: 10.1016/j.anifeedsci.2008.11.003

[95] Hendricksen RE, Gazzola C, Reich MM, Roberton RF, Reid DJ, Hill RA. Using molasses as an alternative to controlled release devices for administering n-alkane markers to cattle. Anim. Sci. 2003;76:471–480.

[96] Elwert C, Kluth H, Rodehutscord M. Effect of variable intake of alfalfa and wheat on faecal alkane recoveries and estimates of roughage intake in sheep. J. Agric. Sci. 2004;14:213–223. DOI:10.1017/S0021859604004150

[97] Smith DG, Mayes RW, Hollands T, Cuddeford D, Yule HH, Malo Ladrero CM, Gillen E. Validating the alkane pair technique to estimate dry matter intake in equids. J. Agric. Sci. 2007;145:273–281. DOI:10.1017/S0021859607006788

[98] Dove H, Mayes RW, Lamb CS, Ellis KJ. Factors influencing the release rate of alkanes from an intra-ruminal, controlled-release device, and the resultant accuracy of intake estimation in sheep. Aust. J. Agric. Res. 2002;53:681–696. DOI:10.1071/AR01175

[99] Ordakowski AL, Kronfeld DS, Holland JL, Hargreaves BJ, Gay LS, Harris PA, Dove H, Sklan D. Alkanes as internal markers to estimate digestibility of hay or hay plus concentrate diets in horses. J. Anim. Sci. 2001;79:1516–1522. DOI:/2001.7961516x

[100] Stevens DM, van Ryssen JBJ, Marais JP. The use of n-alkane markers to estimate the intake and apparent digestibility of ryegrass and Kikuyu by horses. S. Afr. J. Anim. Sci. 2002;32:50–56. DOI:10.4314/sajas.v32i1.3791

[101] Peiretti PG, Meineri G, Miraglia N, Mucciarelli M, Bergero D. Intake and apparent digestibility of hay or hay plus concentrate diets determined in horses by the total collection of feces and n-alkanes as internal markers. Livest. Sci. 2006;100:189–194. DOI:10.1016/j.livprodsci.2005.08.016

[102] López López C, Celaya R, Rodrigues MAM, Osoro K, Ferreira LMM. Combination of long-chain alcohols and fatty acids with alkanes as faecal markers to estimate feed intake and digestibility of horses and cattle fed on grass-heathland vegetation communities. Can. J. Anim. Sci. 2016;96:221–231. DOI:10.1139/cjas-2015-0071

[103] Wilson H, Sinclair AG, DeB Hovell FD, Mayes RW, Edwards SA. Validation of the n-alkane technique for measuring herbage intake in sows. In: Proceedings of the British Society of Animal Science; March 1999; Penicuik, Midlothian, UK: BSAS; 1999. p. 177.

[104] Sehested J, Breinhild KK, Søegaard K, Vognsen L, Hansen HH, Fernández JA, Danielsen V, Kristensen VF. Use of n-alkanes to estimate grass intake and digestibility in sows. In: Dove H, Coleman SW, editors. Nutritional Ecology of Herbivores. Satellite Symposium: Emerging Techniques for Studying the Nutrition of Free Ranging Herbivores. San Antonio, TX (CD-ROM)

[105] Hulbert IA, Iason GR, Mayes RW. The flexibility of an intermediate feeder: Dietary selection by mountain hares measured using faecal n-alkanes. Oecologia 2001;129:197–205. DOI:10.1007/s004420100725

[106] Hatt JM, Mayes RW, Clauss C, Lechner-Doll M. Use of artificially applied *n*-alkanes as markers for the estimation of digestibility, food selection and intake in pigeons (*Columba livia*). Anim. Feed Sci. Technol. 2001;94:65–76. DOI: 10.1016/S0377-8401(01)00294-2

[107] Mayes RW, Lamb CS, Colgrove PM. Digestion and metabolism of dosed even-chain and herbage odd-chain n-alkanes in sheep. In: Proceedings of the 12th General Meeting of the European Grassland Federation; 4–7 July 1988; Dublin: EGF; 1988. pp. 159–163.

[108] Ohajuruka OA, Palmquist DL. Evaluation of n-alkanes as digesta markers in dairy cows. J. Anim. Sci. 1991;69:1726–1732. DOI:/1991.6941726x

[109] Gudmundsson O, Thohallsdottir AG. Evaluation of n-alkanes for intake and digestibility determination in horses. In: Gibb MJ, editor. Proceedings of the IXth European Intake Workshop; 18–20 November 1998; North Wyke, Devon, UK: IGER; 1998. pp. 1–4

[110] O'Keefe NM, McMeniman NP. The recovery of natural and dosed n-alkanes from the horse. Anim. Prod. Aust. 1998;22:337. Available from: http://www.asap.asn.au/livestocklibrary/1998/O'Keefe98a.PDF [Accessed: 2016-07-12]

[111] Vouzela CFM. Validation of n-alkanes as markers in herbivore [thesis]. Angra do Heroísmo: University of Azores; 2002.

[112] Elwert C, Dove H, Rodehutscord M. Effect of roughage species consumed on faecal alkane recovery in sheep, and effect of sample drying treatment on alkane concentrations. Aust. J. Exp. Agric. 2006;46:771–776. DOI:10.1071/EA05303

[113] Himmelsbach DS. Structure of forage cell walls-session synopsis. In: Jung HG, Buxton DR, Hatfield RD, Ralph J, editors. Forage Cell Wall Structure and Digestibility. Madison, WI: ASA-CSA-SSSA; 1993. pp. 271–283. DOI: 10.2134/1993.foragecellwall.c11

[114] Kelman W, Bugalho M, Dove H. Cuticular wax alkanes and alcohols used as markers to estimate diet composition of sheep (*Ovis aries*). Biochem. Syst. Ecol. 2003;31:919–927. DOI:10.1016/S0305-1978(03)00081-4

[115] Vargas-Jurado N, Tanner AE, Blevins SR, McNair HM, Mayes RW, Lewis RM. Long-chain alcohols did not improve predictions of the composition of tall fescue and red clover mixtures over n-alkanes alone. Grass Forage Sci. 2015;70:499–506. DOI:10.1111/gfs.12134

[116] Ferreira LMM, Oliván M, Celaya R, Garcia U, Rodrigues MAM, Osoro K. The use of n-alkanes to estimate diet composition of ruminants grazing on species diverse plant communities—Effect of feeding selectivity on diet composition estimates. Livest. Sci. 2007;111:114–123. DOI: 10.1016/j.livsci.2006.12.008

Chapter 4

Simple Models for Describing Ruminant Herbivory

Kenneth J. Moore, Sotirios V. Archontoulis and
Andrew W. Lenssen

Additional information is available at the end of the chapter

http://dx.doi.org/10.5772/67342

Abstract

The use of quantitative independent variables in experiments allows the use of regression to explore the functional relationship between treatments applied and measured responses. It provides the opportunity to not only understand the magnitude and importance of the response but also ascertain its nature. The simplest approach is to fit a polynomial. While it is often possible to obtain a very good fit using this approach, it offers in the way of providing insight into the response. At best, you can determine if the response is nonlinear and if so, if it is complex or not. The model parameters are empirical and generally cannot be interpreted as having any biological, chemical, or physical meaning—at least not directly. There are situations, however, when such a meaning can be inferred from a model fit using simple regression. In general, this is true when the relationship is truly linear or when a nonlinear model can be considered to be "intrinsically" linear; that is, it can be linearized by transforming the data in a way that can be fit using simple linear regression. A series of forage quality examples are used to illustrate these concepts in this article.

Keywords: modeling, ruminant, herbage quality, digestion, kinetics, true digestibility, intake

1. Introduction

The use of mathematical models to describe chemical, physical, and biological processes is quite common in natural sciences [1]. The best models are those with parameters that have chemical, physical, or biological meanings [2]. They go beyond being descriptive and provide a deeper understanding of the process that is being evaluated. Fitting and adapting models to experimental data are as much an art as a science, and the outcome is highly influenced by decisions made by the researcher about which models to fit, what data are needed and should be used, and interpretation of fit statistics. In this article, models that describe the relationship between rumen escape protein and protein concentration, kinetics of fiber digestion, true digestion, and potential intake of herbage are developed and used to demonstrate how relatively simple models can be an effective tool for understanding biological processes and how they can be applied using experimental data. For each example, the underlying theory and assumptions are also presented and discussed.

It is important to make the distinction between the use of models for describing and understanding a biological response and their use to predict future outcomes. The use of models addressed herein relates to the former purpose and is most applicable to interpreting the results of designed experiments. That is, the experimental units on which the observations are made have been intentionally manipulated in some way that can be described quantitatively. Treatment responses for any experiment for which a quantitative treatment has been applied can be evaluated in this way. However, the inferences that can be made by using models fit to experimental data are limited to those appropriate to the design of the experiment. Their application to predicting results outside the bounds of the inference space associated with the experiment is not recommended.

Designed experiments often have unique features that both limit and extend the types of regression analyses that can be performed. They nearly always include multiple replications of individual treatments. When fitting a regression equation, this allows for the partitioning of residual error into pure error and lack of fit, thus providing a test for whether the linear model fits the response or not. It also allows statistical tests to be made about assumptions related to the distribution and homogeneity of residuals. These assessments can be used to refine the approach used in the regression and improve the value of the analysis.

This chapter is intended to demonstrate how relatively simple models can be used to describe important nutritional processes related to ruminant herbivory. It uses a series of examples to illustrate the principles and power of using simple mathematical models to better understand the functional relationship between important variables. The data used in the examples have been published previously, although the analyses employed here may be slightly different from those used in the original studies from which they were taken.

2. Linear regression; a quick review/overview

Simple linear regression is a statistical method for calculating parameters for the model:

$$\hat{Y} = b_0 + b_1 X \quad (1)$$

Graphically, the model represents a straight line that intercepts the Y axis at b_0 and which has a slope equal to b_1. As X increases, Y either increases or decreases proportionally depending on whether the slope is positive or negative, respectively. The model parameters, b_0 and b_1, can be estimated using least squares regression. This approach is based on an algebraic solution of normal equations and produces parameters that minimize the sum of the squared deviations

http://dx.doi.org/10.5772/67342

of observed values from those predicted by Eq. (1). The regression line intersects the point that represents the means of X and Y unless it has been forced through the origin ($X = 0$, $Y = 0$) and the sum of the deviations from the regression line is zero.

The equation for estimating the slope (b_1) is as follows:

$$b_1 = \frac{\sum(X_i - \bar{X})(Y_i - \bar{Y})}{\sum(X_i - \bar{X})^2} \tag{2}$$

Once b_1 is known, then b_0 can be estimated using the equation:

$$b_0 = \bar{Y} - b_1\bar{X} \tag{3}$$

The assumptions for linear regression are the following: (1) the independent variable X is measured without error; (2) the relationship between X and Y is linear; (3) deviations from the regression are independent; (4) the variance in Y is homogenous or constant across the range in X; and (5) the residuals or deviations from the regression are distributed normally. There are ways to assess whether most of these assumptions are valid or not, and they will be described where appropriate in the examples that follow.

Some straightforward statistics for assessing the fit of a regression equation are the coefficient of determination (r^2) and the standard error of the estimate ($S_{Y \cdot X}$). The coefficient of determination is calculated as follows:

$$r^2 = \frac{\sum\left(\hat{Y}_i - \bar{Y}\right)^2}{\sum(Y_i - \bar{Y})^2} \tag{4}$$

It represents the proportion of total variation in Y explained by the regression model and varies between 0 and 1. A value approaching 1 indicates that the regression equation explains most of the variation in Y and, therefore, does a good job explaining the relationship between Y and X. The coefficient of determination is the square of the simple correlation coefficient (r) that is interpreted as the degree to which X and Y vary together. The correlation coefficient is used to describe the relationship and varies from −1 to +1 indicating whether Y decreases or increases with respect to X, respectively. Values close to −1 or +1 indicate a high degree of association between Y and X. The simple correlation coefficient can be calculated as the covariance between X and Y divided by the square root of the product of the standard deviations in X and Y and thus can be thought of as a standardized covariance.

The standard error of the estimate is calculated from the equation:

$$S_{Y \cdot X} = \sqrt{\frac{\sum(Y - \hat{Y})^2}{n-2}} \tag{5}$$

It is the square root of the residual variance of the regression. It describes how well the regression line fits the data with smaller values indicating a better fit. Smaller values indicate less departure of the actual observations from the regression line.

These five equations are all that is needed to fit and assess models that are either linear or intrinsically linear. However, there are other methods and statistics that are useful for this purpose and some of them will be described as the examples that follow are developed.

3. Intrinsically linear models

Nonlinear models that can be linearized by transforming either *Y* or *X* are considered to be intrinsically linear and can be fit using simple linear regression [3]. The most common of these involves logarithmic transformations of *X* or *Y* to yield a linear model with two parameters: (1) exponential (log-linear), (2) logarithmic (linear-log), and power (log-log) functions.

The exponential model has a number of important uses. With a positive slope (b_1), it can be used to describe exponential growth that is unbounded. With a negative slope, it can be used to describe exponential decay. In this form, it is useful in isotope studies to describe radioactive decay and is used in marker dilution studies in a similar manner. It can be also used to describe first-order kinetics for chemical reactions. We will use it in this latter context in an example on modeling herbage digestion that follows.

The logarithmic and power models are useful for describing responses where the rate of change gradually decreases with respect to increasing X. Many chemical and biological processes are limited and show an asymptotic response. These types of responses are generally better described by an intrinsically nonlinear model that contains an asymptote as a parameter. The Gompertz [4] and Mitscherlich [5] equations are two good examples of such models commonly used to describe biological processes. However, these models require a different approach to estimating their parameters than simple linear regression.

If a nonlinear model cannot be expressed in the form of a simple linear equation through transformation, then it is considered to be intrinsically nonlinear. There is a host of such models, and many of them can be used to describe functional responses relevant to herbivory (see Archontoulis and Miguez, 2013, for a review of 77 nonlinear models). However, fitting these models is somewhat more complicated and requires using a numerical approach that adjusts parameter values iteratively until a solution based on certain criteria is achieved. The criterion typically used is the combination of parameter estimates that results in the minimum residual sum of squares, which is why such algorithms are sometimes referred to as a nonlinear least squares approach [6]. Convergence is then based on identifying the combination of parameter estimates that result in the lowest sum of squared deviations from the value estimated by the regression. This chapter focuses on models for which the parameters can be estimated by simple linear regression all of which are therefore intrinsically linear.

4. Herbage nutritional entities

The forgoing concepts and equations can be used to fit and assess simple linear regression equations. However, before applying them to experimental data in the following examples, a quick overview of some nutritional concepts related to herbage utilization is in order.

The nutritive entities of herbages are broadly grouped into uniform and nonuniform fractions based on the Lucas Test [7]. Uniform fractions are those that have similar nutritional characteristics or true digestibility regardless of the feedstuff [8, 9]. These include most nutrients contained in the cytoplasm of plant cells including proteins and other nitrogenous compounds and nonstructural carbohydrates. Nonuniform fractions vary in true digestibility among different feedstuffs and even within a single feedstuff. Plant fiber is considered a nonuniform fraction. Its digestibility varies greatly among different feedstuffs and is affected by a number of genetic and environmental factors [10].

The Lucas test itself involves a simple linear regression model. It is performed by regressing the amount of a nutrient that is digestible against its intake. Fractions for which true digestibility is constant over a range of herbages are considered to be nutritionally uniform or ideal [10]. The Lucas Test provided the foundation on which Van Soest [8] developed the detergent system for analyzing feeds. In this system, herbage or feedstuff dry matter is partitioned into cell solubles and neutral detergent fiber by refluxing a sample of the feed in a neutral detergent solution and recovering the residue by filtration. The residue remaining is fiber. The compounds removed with the filtrate are collectively referred to as cell solubles. Cell solubles have a uniformly high true digestibility regardless of the feedstuff they are contained within. They are very nearly completely available when subjected to digestion in ruminants with a true digestion coefficient of 0.98. The residue remaining after treatment with neutral detergent is the fibrous fraction and varies significantly in digestibility among feedstuffs. Both fiber and cell solubles are heterogeneous in composition and can be further partitioned into chemical constituents. Neutral detergent fiber, while structurally complex, is composed of relatively few polymers and consists almost entirely of cellulose, hemicelluloses, and lignin. True, the hemicelluloses represent a fairly diverse set of compounds, but still this is a relatively small number compared with the myriad of compounds found in plant cells. Some complex carbohydrates such as pectins and beta glucans are not recovered in neutral detergent fiber. However, these compounds are easily digested by ruminants and are considered to be part of the cell soluble fraction [11].

A particularly important fraction of the cell soluble fraction is protein. Proteins and other nitrogenous compounds in herbage can be converted to amino acids by rumen microorganisms that incorporate them into proteins. These proteins are eventually passed from the rumen to the lower digestive tract where they are hydrolyzed to amino acids, which are largely absorbed within the small intestine [12]. The example that follows involves using a modification of the Lucas Test to test the hypothesis that rumen degradability of protein is proportional to the concentration of protein in the herbage.

5. Rumen degradable protein

In this application, a modified form of the Lucas Test is used to evaluate the degradability of herbage in the rumen using an *in situ* technique. The data are from an experiment designed to assess ruminal degradation of smooth bromegrass (*Bromus inermis* Leyss.) and switchgrass (*Panicum virgatum* L.) using an *in situ* analytical technique [13].

The degradability of protein in the rumen varies greatly between cool and warm-season grasses, and this may be one explanation for the observation that animals consuming warm-season grasses perform better than would be expected based on their chemical composition.

The theory is that some plant proteins localized within bundle sheath cells in warm-season grasses are physically protected from degradation by the structure of the cells. These proteins bypass the rumen intact and progress to the lower intestinal tract where they are digested and absorbed as amino acids. Nitrogen in proteins degraded in the rumen is often in excess of microbial needs and that which is not needed is lost as ammonia. Thus, protecting some of the protein from ruminal degradation improves the efficiency of protein utilization [12].

One of the objectives of the experiment was to quantify the relationship between ruminal protein degradation and protein concentration for both species. Linear regression of rumen degradable protein (RDP) on crude protein (CP) concentration was done for several samples of both species with varying CP concentration. The linear equation for this analysis was:

$$\mathrm{RDP} = d_0 + d_1\,\mathrm{CP} \tag{6}$$

where d_0 represents the endogenous contribution to RDP and d_1 is the true digestion coefficient for ruminal degradability.

The grasses used in this study were harvested at different stages of maturity and separated into leaves and stems to obtain a range of CP concentrations and were analyzed for RDP using an *in situ* bag technique. Samples of each grass were enclosed in Dacron bags and incubated in the rumen of a live animal for 12 h. The loss of protein from the bag was determined by difference using residual protein remaining after digestion, and RDP was calculated based on protein disappearance from the bag. The endogenous contribution (d_0) in this system represents microbial contributions of protein to the residue remaining after incubation by rumen microbes.

Model parameters for each of the two grass species were estimated using the REG procedure in SAS (Appendix A1). The linear model described the relationship between RDP and CP very well for both species (**Figure 1**) based on the r^2 and standard error of the estimate (RMSE in SAS output). The calculated coefficient for ruminal degradability of CP was 74% for smooth bromegrass and 57% for switchgrass. The null hypothesis that these two slopes are the same can be tested with a *t*-test:

$$t = \frac{b_1 - b_2}{S_{b_1 - b_2}} = \frac{0.739 - 0.572}{0.058} = 2.88 > 2.09\ t_{0.05,19df} \tag{7}$$

where

$$S_{b_1-b_2} = \sqrt{\frac{(S^2{}_{Y\cdot X})_p}{\sum(X-\bar{X})_1^2} + \frac{(S^2{}_{Y\cdot X})_p}{\sum(X-\bar{X})_2^2}} = \sqrt{\frac{38.115}{32499.27} + \frac{38.115}{17466.65}} = 0.058 \tag{8}$$

$$(S^2{}_{Y\cdot X})_p = \frac{(residual\ SS)_1 + (residual\ SS)_2}{(residual\ DF)_1 + (residual\ DF)_2} = \frac{484.279 + 239.903}{9 + 10} = 38.115 \tag{9}$$

Based on this comparison, it is reasonable to conclude that the two species have different rumen protein degradability and that this difference is constant and persists across a range of maturities and morphological components. These results are consistent with the observation that protein in warm-season grasses seems to be used more efficiently than that in cool-season grasses. However, the mechanism for why this is so is not clear from this study. Based on the protection theory, one might expect protein degradability to vary across tissues within a species and that does not appear to be the case. So maybe there is another explanation that would better describe what was observed in this study.

http://dx.doi.org/10.5772/67342

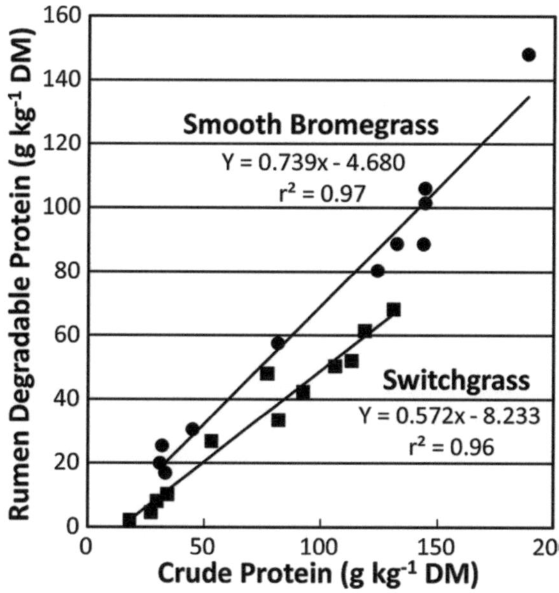

Figure 1. Relationship between rumen degradable protein and protein concentration in smooth bromegrass (●) and switchgrass (■). Individual data points represent leaf and stem samples collected at different stages of maturity. The slope of the equation represents the proportion of the dry matter that is degraded in the rumen and the intercept can be interpreted as the microbial contribution. Data from Ref. [13].

One further conclusion that can be inferred from fitting these equations is that the contribution of microbial CP to residual CP was negligible for smooth bromegrass and small (<1%), but significant for switchgrass. This is based on the test of parameter estimates included in the SAS output. The *t*-test for the intercept associated with the smooth bromegrass model was not significantly different from zero ($P > 0.05$), while that for the switchgrass model was ($P < 0.05$).

6. Fiber digestion kinetics

In this application, we will compare the digestion of alfalfa (*Medicago sativa* L.) and tall fescue (*Festuca arundinacea* Schreb.) by fitting a log-linear model to calculate the rate constant and lag time (Appendix A2). The data are from an experiment that was designed to compare different approaches for estimating the parameters of a first-order digestion model [14]. In this case, we will be using the data to explore some unique aspects of such data sets and how they require some rearrangement and culling of data in order to fit a first-order model to them. Once this is done, we will use simple linear regression to calculate the rate of digestion and then estimate a lag period based on the intercept of the equation.

To comprehend and be able to interpret the parameters of the model, an understanding of plant fiber and first-order kinetics is necessary. The next two sections provide an overview of each of these topics following that we will pick up the example in more detail.

6.1. Plant fiber

Fiber is a nutritional concept that refers to the less degradable and more variable constituents of an herbage or feedstuff. Chemically, it is comprised of plant cell walls the composition of which varies greatly among and within herbage species. Even within a single plant, the organs, cells, and tissues vary remarkably in fiber composition and digestibility [15]. The primary chemical constituents of plant fiber are cellulose, hemicellulose, and lignin, although there are others that comprise a much smaller fraction. These constituents are aggregated and arrayed in three-dimensional space in various ways creating a network of nonliving tissues that play an important structural role in the architecture of their plant [11]. Having a rigid cell wall is one of the defining characteristics of higher plants. Cell walls and thus fiber evolved to fulfill specific roles in plants, which do not include being a source of energy for herbivores. Their structure and function in a plant are in many ways counter to their use as a nutrient source. Even though fiber is composed of plant cell walls, it is functionally different. It is defined by its properties when subjected to digestion by an animal and has attributes that are only relevant in this context. The two terms are thus not really interchangeable [11].

Not all the fiber in a plant is degradable by ruminants. The degradation of plant fiber involves the hydrolysis of the principal polysaccharides by enzymes secreted by rumen bacteria. Because of the close physical and chemical interactions among plant cell wall constituents, some of the glycosidic linkages are not accessible to the hydrolases that would otherwise cleave them and render them digestible. The fraction of fiber that cannot be digested because of these interactions is indigestible and cannot be degraded within the digestive system. When determining the kinetics of fiber digestion, the indigestible (C_I) portion must be considered separately and removed from that which is potentially digestible (C_D) [16, 17].

The indigestible fraction is usually considered to be that which remains after being subjected to *in vitro* or *in situ* digestion for a period of time. This is usually between 48 and 96 hours and well past the expected residence time that it would be exposed to digestion in the rumen of an animal. The potentially digestible fraction is the difference between total fiber (C_0) and indigestible fiber ($[C_D]_t = [C_D]_0 - [C_I]$), and its concentration decreases exponentially during digestion asymptotically approaching zero according to the first-order rate law. The key to defining the indigestible fraction is to subject the herbage to digestion long enough to approach an asymptote after which time no further digestion occurs (**Figure 1**). Once this is achieved, there is very little change in the concentration digested. Indigestible fiber is usually calculated as the concentration of fiber remaining after incubation for 72–96 h *in vitro*. The period needed to reach this point varies with substrate and periods as short as 24–36 h for rapidly degraded herbages are not uncommon.

When calculating first-order digestion parameters, it is important only to include fiber concentrations at time points where digestion is actively occurring ($[C_D]_t \neq [C_D]_0$ and $[C_D]_t \neq [C_I]$; $[C_D]_0 > [C_D]_t > [C_I]$). Including time intervals in the calculation where no change in fiber concentration has occurred biases the estimates of the parameters. Most importantly, time intervals where the fiber concentration is not different from either the initial or final concentration should be excluded from the calculations. Moore and Cherney [13] suggested a simple method for selecting time intervals for rate calculations. Since replicate samples are usually collected at each time point during a digestion study, it is possible to compare the mean concentration between pairs of time points using a t-test. Time intervals within the lag period can be identified as those for which the fiber concentration is not significantly different than the initial concentration. Time intervals occurring after digestion has ceased will have concentrations that are not different than the longest time point, which is usually used to determine the indigestible fiber fraction. It is entirely possible for digestion to occur throughout the sampled period, but it is more often the case that some time points will need to be excluded.

The data in Appendix 2 were collected by incubating an herbage sample in buffered rumen fluid for 0, 3, 6, 9, 12, 16, 24, 36, 48, 60, 72, and 96 h [14]. Samples were refluxed in a neutral detergent solution following fermentation to extract undegraded fiber using the procedures described by Cherney et al. [18]. The concentration of fiber remaining after each time interval was calculated on the basis of initial dry matter subjected to fermentation (**Figure 2**).

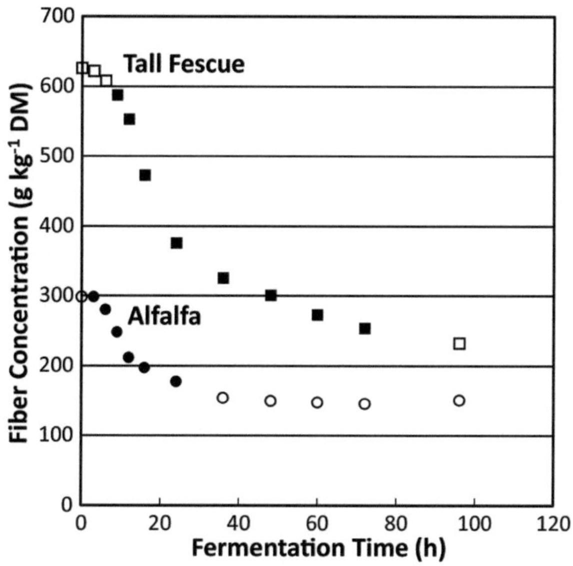

Figure 2. Concentration of fiber remaining during fermentation of alfalfa (●) and tall fescue (■) samples incubated in buffered rumen fluid. Symbols represent means of four subsamples and those that are closed were determined to be different to adjacent means using a least significant difference at alpha 0.05. Data from Ref. [14].

In order to calculate an unbiased estimate of digestion rate (*k*), time intervals where no change in concentration occurred must be excluded from the dataset. This was accomplished using the method referenced above using the SAS code presented in Appendix A2.1. This involved conducting a one-way analysis of variance followed by a post-hoc mean comparison using a least significant difference (LSD). Because the measurement represented a wide range in concentrations and the concentrations were quite small after longer time intervals, a test of homogeneity was performed to validate the use of a pooled error for calculating the LSD. Abridged versions of the results that include output from the homogeneity test and LSD are presented in Appendix A2.1.

Based on the Levene test for homogeneity, the variances observed in neutral detergent fiber (NDF) concentration were homogeneous across each time interval and this was true for both species. This simply confirms that using the LSD procedure for comparing the mean NDF concentration remaining at each time interval was appropriate. Had they been found to be heterogeneous, it would not have been possible to use a pooled estimate of the variance for comparing means and either a data transformation to stabilize the variance or the use of different variances for each comparison would be needed. There are other tests of homogeneity that can be used to assess whether treatment variances can be considered equal. The Levene test is the default method when using GLM in SAS because it is widely used and accepted, but others are available and can be specified if desired.

The mean concentration associated with each time interval is reported in the output along with a capital letter denoting which grouping it belongs to (Appendix A2.1). Means associated with the same letter are not different from each other at alpha 0.05. By comparing the groupings, it is possible to infer the intervals during which digestion began and when it ceased. Digestion of alfalfa fiber began after the first interval so the concentration at 3 h would be included in the calculation of *k* (**Figure 2**). However, for tall fescue digestion did not begin until sometime between 6 and 9 h so the concentrations at 3 and 6 h would be excluded (**Figure 2**). Digestion of alfalfa fiber ceased after 24 h as there was no difference between the concentrations at this time and the next one at 36 h. For tall fescue, fiber digestion continued throughout the remainder of the incubation period once it began.

This analysis (Appendix A2.1) can also be used to determine the concentration of indigestible fiber that must be subtracted from the concentration ($[C]_t$) remaining after each time interval (**Figure 2**). The longest time period is often selected based on the assumption that no further digestion will occur beyond it and the concentration remaining at that time is considered to be indigestible and is used to define $[C_I]$. In Example 2.1, $[C]_{96}$ might be selected to define this fraction. However, instead we will use the mean of the concentrations that were determined to be not different after the end of the period of active digestion. For alfalfa, this will be the concentrations for intervals 48–96 h and $[C]_I$ is then 148.2 g kg^{-1} DM. The 36-h interval is excluded because the *t*-test was ambiguous about which group it belonged to. Since digestion continued through 96 h for tall fescue, the concentration at that time, 232.3 g kg^{-1} DM, was used to define $[C_I]$.

The concentration of digestible fiber is the initial fiber concentration minus the indigestible fraction ($[C]_0 - [C_I]$) which in this study was 153.1 g kg^{-1} DM for alfalfa and 393.8 g kg^{-1} DM for tall fescue (**Table 1**). This is the fraction or pool to which first-order kinetics applies. To calculate *k*, the rate of fiber digestion, C_I must be subtracted from each observed value of *C* that occurred during the incubation. Perhaps a better way of describing this is to say that indigestible portion of the residue remaining after each time interval is constant and must be subtracted from the total amount remaining for first-order kinetics to apply.

6.2. First-order model

The rate of fiber digestion in the rumen is dependent on the concentration present and, therefore, follows first-order kinetics:

$$\frac{d[C_D]}{dt} = k[C_D] \tag{10}$$

where $[C_D]$ is the digestible fiber concentration remaining at time *t* and *k* is the first-order rate constant.

Parameter/quantity	Alfalfa	Tall fescue
Fiber $[C]_0$, g kg^{-1} DM	298.3	626.0
Indigestible fiber $[C_I]$, g kg^{-1} DM	148.2	232.3
Digestible fiber $[C_D]$, g kg^{-1} DM	150.1	393.8
Cell solubles $[C_S]$, g kg^{-1} DM	701.7	374.0
Rate of fiber digestion k, h^{-1}	−0.111	−0.044
Lag time, h	1.7	5.3
C_D/C_0	0.503	0.629
True digestibility, g kg^{-1} DM	851.8	767.8

Table 1. Model parameters calculated for fiber digestion and true digestion of alfalfa and tall fescue (Example 2).

The absolute rate of change in concentration per segment of time depends on the concentration present during that segment. Higher rates of digestion for a given herbage substrate correspond to higher concentrations since the relative rate (k) is constant. So even though the absolute rate changes during fermentation, the proportional rate at which C is degraded is constant throughout. The units of the rate constant are in reciprocal time, which for plant fiber is generally measured in hours (h^{-1}). Because k is a proportional constant, it can be expressed as a percentage as well as a fraction and is sometimes given in percentage units.

Eq. (10) can be written in differential form and divided by C_D to give the equation:

$$\frac{d[C_D]}{[C_D]} = -kdt \tag{11}$$

Integrating both sides gives:

$$\ln[C_D] = -kt + C \tag{12}$$

where C is the constant of integration.

This is a convenient form of the equation because the parameters can be calculated by simple linear regression of the logarithm of concentration remaining over time. This is the exponential decay model referred to in the discussion of intrinsically linear models above. Concentrations in Eq. (12) are in log units. C is the constant of integration and in this context is the logarithm of digestible fiber concentration (C_D) in the absence of a lag period. It is the intercept where the regression line intersects the ordinate at $t = 0$. Because a lag period is observed for most *in vitro* incubations, the value given by C is rarely equal to the logarithm of initial digestible fiber concentration. Whenever a lag period occurs, the concentration at $t = 0$ will be greater than the known concentration in the substrate.

The rate constant (k) and lag time (L) can be calculated using linear regression on data collected from *in vitro* or *in situ* incubations. Samples of herbage are incubated for several time intervals, and the concentration of fiber remaining is determined for each one. Since kinetics apply only to the fraction of fiber that is potentially digestible, the fraction that is indigestible must be determined and subtracted from that remaining at each time interval. In Example 2, the indigestible fiber concentration was calculated as the mean concentration remaining after digestion had ceased and was determined by using a mean comparison test to determine the time interval after which no further decreases in fiber concentration occurred (Appendix A2.1). The natural logarithm of the fiber concentration remaining ($[C]_t$) minus the indigestible fiber concentration ($[C_I]$) during this period of active digestion is shown for both species in **Figure 3**. The symbols in the figure represent the mean of the four replicate measurements of fiber concentration that were made at each time point. Regression of $\ln([C]_t - [C_I])$ on time using individual data points (i.e., replicate data) or mean at each *X* (i.e., time) results in the same estimates of the parameters for the linear model (Eq. (12)), although the fit statistics will vary. However, since there are multiple values of *Y* for each *X*, it is possible to partition the residual sum of squares (SS) into lack of fit and pure error [19]. Theoretically, if the model fits, then squared deviations of each observation from the regression and squared deviations from the mean of each of *X* value should be the same. The difference between these two values is called lack of fit and its significance can be assessed using an *F*-test. If the lack of fit variance is determined to not be different from that for pure error, then it can be concluded that the correct model was used to describe the relationship between *Y* and *X*. For both alfalfa and tall fescue, there was no lack of fit of the linear model and we can be confident in the parameters that were estimated by the regression (Appendix A2.2).

There are also qualitative methods for assessing the appropriateness of a model for a set of data. A residual is defined as the difference between the observed value and that estimated by the regression. Examination of residuals is a fast and easy way to visualize the fit of a model and determine if it is biased for some values of *X*. Ideally, the residuals will appear to be randomly distributed around the regression line with no obvious clustering above or below along any segment of *X*. This can be observed for the example in **Figure 3** that displays the mean digestible fiber concentration at each time point and regression lines for both species. The residual at each *X* is the distance between the symbol and the line. A plot of actual residuals is a common output of statistical analysis programs, but the patterns are easily visualized when the regression line is plotted with the means as shown in **Figure 3**. In this case, no patterns are discernible in the residuals for either species, and it is readily evident that the linear model fits the data.

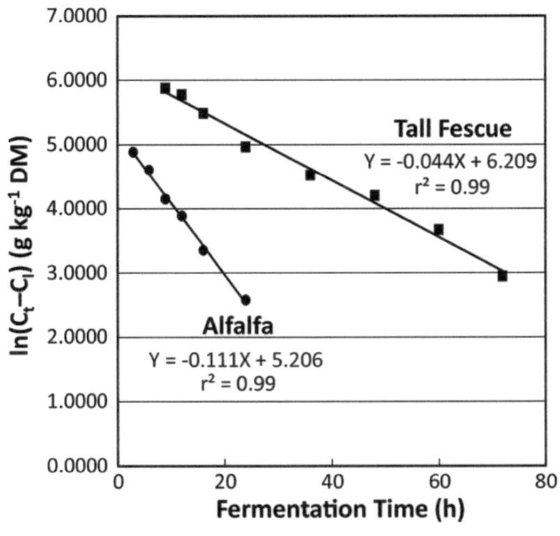

Figure 3. Regression of the natural logarithm of digestible fiber on time of fermentation for alfalfa (●) and tall fescue (■) samples. Only data from within the period of active digestion were included in the calculation. The slope of the line represents the first-order rate of digestion. Data from [14].

The constant for rate of fiber digestion was 0.111 h^{-1} for alfalfa and 0.044 h^{-1} for tall fescue. These two slopes can be compared using a *t*-test similar to that used in the first example:

$$t = \frac{b_1 - b_2}{S_{b_1 - b_2}} = \frac{0.111 - 0.044}{0.0156} = 4.30 > 2.01\ t_{0.05,52df} \quad (13)$$

Based on this comparison, it is clear that the rate of fiber digestion is quite different between the two species with the digestion of alfalfa fiber occurring at over 2.5 times the rate as that of tall fescue. For reasons we will see, interpreting the rate of fiber digestion independently of concentration and degradability can be misleading. However, at this point, for whatever reasons, we conclude that in this study the rate at which fiber was digested in alfalfa was much faster than in tall fescue.

It is possible to test multisample hypotheses about slopes when more than two equations are being compared; for example, if an additional species were being considered in this example. This can be accomplished using analysis of covariance [20], but the test is not easily implemented using statistical software. The calculations involved are laborious enough to consider avoiding making them altogether and instead making pairwise comparisons between slopes using the *t*-test described above. In many cases, as in this example, the slopes are clearly different, and it would be reasonable to proceed with interpreting what the differences mean under that assumption. There are other ways to convince yourself and others that this is appropriate. For example, a significant interaction between species and time in an analysis of variance of digestible fiber concentration remaining over time indicates that the slope of the response is different between species. Even when the slopes are the same, it does not mean that the equations are the same. The intercept can differ between two simple linear equations with the same slope. There are again statistical tests that may be used to compare the intercepts of linear equations [20], but in this case, it is far easier to conduct an analysis of variance or *t*-test on the initial digestible fiber concentrations (**Table 1**).

It is common for there to be a lag period before fiber digestion begins in *in vitro* digestion systems [21]. This is usually attributed to the time required by the rumen bacteria to colonize the sample and begin growing in number. There is often no measurable lag time in *in situ* digestions systems [22]. Sometimes, a negative lag time is observed *in situ* that is attributed to washout of particles from the bag containing the sample and is an artifact of the method. Whether or not there is an actual lag time before digestion begins in the rumen is a subject of debate. It is most likely a function of the substrate and depends on the chemical and physical attributes of the plant material. Longer lag times have been observed for more mature herbage, suggesting that lignification may play a role in delaying active digestion [14]. It takes time for fibrolytic bacteria to colonize herbage particles and for hydrolytic enzymes to access their target polysaccharides. It is reasonable to assume that these processes occur faster for herbage consumed naturally. Regardless, when a lag period occurs during an incubation in which the rate of digestion is to be determined, it must be accounted for in the calculations.

The lag time (L) can be calculated by determining the time when C is equal to the initial concentration of digestible fiber:

$$L = \frac{(\ln [C_D] - C)}{k} \quad (14)$$

That is L is equal to time required to digest the difference between the actual digestible fiber concentration and that predicted by the regression. The lag period will be equal to the time required to reach the initial concentration. Lag time is a constant and for this reason is considered discrete. It can be calculated by first solving for k and C in Eq. (12) by linear regression and using these values to solve for L using Eq. (14).

The foregoing discussion raises the question of how many time points are needed to accurately estimate kinetic parameters. Theoretically, the answer is only two as long as the substrate concentrations are measured accurately, concentrations at the two selected time points are different than the initial concentration and final concentration, and the first-order model can be assumed. With only two time points, the calculation of k is simplified:

$$k = \frac{\ln\left(C_{t_1} - C_I\right) - \ln\left(C_{t_2} - C_I\right)}{t_2 - t_1} \quad (15)$$

As long as fiber digestion is occurring throughout the interval defined by the two time points, then k should be constant throughout the entire time period during which digestion occurred. This may be the most practical method for estimating k when the fiber digestion of a large number of treatments is being evaluated as it greatly reduces the number of subsamples that must be incubated.

There are advantages to using more than two points. Each additional time point decreases the leverage of the others. Since it is not possible to measure the fiber concentration with absolute accuracy, the slope of a line computed with only two points may be subject to higher error than one calculated with more points. Because a line is defined by two points, it not possible to assess if the linear model describes the relationship so it is necessary to assume that it does. However, one might argue that it is better to have more replications to estimate a mean concentration for a few values of X rather than several unreplicated values at many values of X. As long as the assumptions for fitting the model hold, it probably does not much matter, although there may be other nonstatistical reasons for using one approach or the other.

7. Fiber digestion model

The parameters k, C_D, and L together describe the digestion kinetics of herbage fiber and can be used to predict the expected concentration of digestible fiber remaining after an interval of digestion:

$$[C_D]_t = [C_D]_0 \, e^{-k(t-L)} \quad (16)$$

This equation is the nonlinear form of Eq. (12) that has been adjusted for L and uses actual units of concentration rather than log units. The parameters of this equation can be calculated using nonlinear regression. However, using that approach can lead to biased estimates of k and L for herbage with long lag periods, so the log-linear approach is preferred [14].

This equation indicates that the digestible fiber concentration at any given time (t) is a function of the initial concentration ($[C_D]$) multiplied by a proportion equal to $e^{-k(t-L)}$. This proportion can be thought of as the inverse of the digestion coefficient for $[C_D]$ at t. By subtracting the concentration of digestible fiber remaining $[C_D]$ predicted by the model from the amount present at time zero $[C_D]_0$, it is possible to calculate the amount digested at any point in time during the incubation:

$$[D]_t = [C_D]_0 - [C_D]_0\, e^{-k(t-L)} \tag{17}$$

which factors to:

$$[D]_t = [C_D]_0 (1 - e^{-k(t-L)}) \tag{18}$$

The units of $[D]_t$ are in concentration per unit of dry matter (e.g., g NDF kg^{-1} DM). Fiber digestibility can be calculated by dividing $[D]_t$ by the concentration of fiber initially present (**Table 1**).

Eq. (18) is one form of a common nonlinear model that is used to describe asymptotic increase. It goes by many names including the monomolecular equation that is used to describe chemical reactions involving a single molecule and the Mitscherlich equation that is used to describe crop yield responses to fertilizer. Archontoulis and Miguez [1] simply refer to it as "Exponential gives rise to maximum." It is useful for characterizing a host of biological relationships that exhibit asymptotic behavior. In its simplest form, it is a two-parameter equation, but a third parameter is sometimes used as a scaling factor to reduce the pool size [23] affected by the proportion defined by $e^{-k(t-L)}$, essentially creating a nonzero intercept on the Y axis. Perhaps more to the point, the monomolecular equation has been used to describe the digestion of herbage [24, 25], although it has been used mostly for characterizing protein degradation which also follows first-order kinetics.

As presented in Eq. (18), the monomolecular equation cannot be linearized in a manner that lends itself to an algebraic solution for all parameters. Estimates of all parameters, however, can be obtained simultaneously using nonlinear regression. In the case of fiber digestion, the parameters are well defined chemically and biologically, and it is more straightforward and, therefore, advantageous to determine their values directly and sequentially as demonstrated in Examples 2.1 and 2.2 (**Table 1**).

Estimates of the digestion model parameters (Eq. (18)) are presented in **Table 1**. We have already concluded that the rate of digestion was over two times as fast for alfalfa as for tall fescue. However, this rate only applies to the digestible portion of fiber, which was much greater in tall fescue. The absolute rate of fiber digestion (Eq. (10)) at the onset of digestion (L) was 16.7 g kg^{-1} DM/h for alfalfa and 17.3 g kg^{-1} DM/h for tall fescue reflecting the much higher concentration of digestible fiber in the latter, which was over two times greater than that for alfalfa (**Figure 4**). Evaluating k only without considering the size of C_D would be very misleading when evaluating the contribution of fiber to the digestible energy available from a particular herbage.

In the next section, we will combine the information we developed in Example 2 with a theory developed by Van Soest and colleagues [8] to estimate the true dry matter digestibility of herbage.

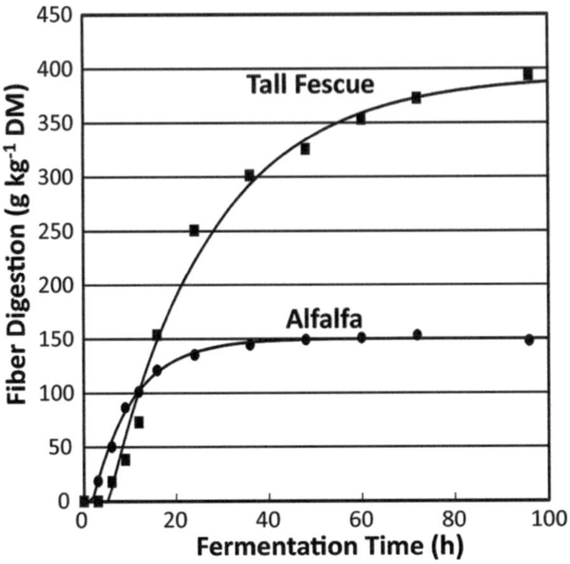

Figure 4. Amount of fiber digested during fermentation for alfalfa (●) and tall fescue (■). The values estimated by the line were calculated using Eq. (18). The first-order rate constant (*k*) was calculated using the subset of data points that were determined to occur during the period of active digestion and the lag time (L) before digestion began was calculated using Eq. (14). Data from Ref. [14].

8. True digestion model

True digestion of herbage dry matter is distinguished from apparent digestion by the contribution of the animal to fecal dry matter [10]. That is, some of the dry matter contained within the feces arises from the animal and microbes inhabiting its digestive system. This includes microbial cells and material sloughed from the walls of the digestive tract [8]. The animal, however, does not contribute fiber to the feces. All the fibers contained within the feces originate from the plant matter consumed by the animal. Given that all dietary constituents other than fiber are virtually digested completely, the true digestibility of the diet can be calculated by excluding nonfiber components from the fecal dry matter. Thus, the true digestibility coefficient of an herbage or diet can be calculated as:

$$TD = \frac{DMI - C_U}{DMI} \tag{19}$$

where DMI is the dry matter intake; the amount of DM consumed and C_U is the undigested fiber; the amount of undigested fiber excreted in the feces.

Depending on the composition of the diet and its physical form, there may be a difference between undigested fiber and that which is truly indigestible [22]. In diets with high passage rates, some of the fiber that may have been digested if exposed to the rumen environment for a longer time escapes undigested and is recovered in the feces. However, for diets that consist entirely or mostly of herbage, it is reasonable to assume that the indigestible fiber fraction reasonably reflects that portion of the diet that is undigested. In *in vitro* and *in situ* assessments of

http://dx.doi.org/10.5772/67342

herbage digestibility, there is no influence of rate of passage and recognizing that dietary and other factors may influence *in vivo* measurements, the true digestibility values obtained from these assessments may be thought of as representing potential values and may not be reached *in vivo* under certain circumstances.

The comprehensive system for feed analysis developed by Van Soest [8] partitions herbage dry matter into two primary fractions based on studies of nutritive uniformity using the Lucas test: cellular contents and neutral detergent fiber. Cellular contents are nutritionally uniform in that they have the same true digestibility across a range of herbages and other feedstuffs. They are virtually completely digestible and according to Van Soest have a true digestion coefficient of 0.98. Digestibility of fiber varies greatly among herbages as we have seen.

Van Soest [8] used these relationships to develop a summative equation for predicting true and apparent digestibility. Accordingly, true digestibility is the sum of cell contents (×0.98) and digestible fiber (C_D). Digestible fiber was estimated either from *in vitro* analyses or a calculation using a simple linear relationship between lignin concentration and fiber digestibility. Apparent digestibility is true digestibility minus 129 g kg^{-1} DM, the latter quantity representing endogenous contributions of nonfiber material to fecal dry matter. Based on these principles, it is possible to estimate true digestibility by summing the output from the fiber digestion model with the contribution from cell solubles:

$$TD = [C_S] + [C_D]_0(1 - e^{-k(t-L)}) \tag{20}$$

where $[C_S]$ is the concentration of cell solubles $(1000-[C]_0)$ [26]. This varies slightly from Van Soest's equation by assuming that cell solubles are completely digestible; i.e., have a digestion coefficient of 1.0 rather than 0.98. For the purposes for which this equation can reasonably be used this variance matters little, but it can be easily corrected by applying Van Soest's digestion coefficient to the cell soluble fraction.

Adding the $[C_S]$ term to the equation essentially changes the *Y* intercept. Since the concentration of cell solubles is completely digestible, at $t < L$, the true digestibility is equal to $[C_S]$ at the beginning of fermentation. Once *t* exceeds *L*, then the proportion of [CD] defined by $(1 - e^{-k(t-L)})$ is added to this amount and true digestibility continues to increase until all of the digestible fiber fraction is digested (**Figure 5**).

Having a model that includes the principal parameters affecting herbage digestion allows assessment of how each entity and the parameters acting upon it influence herbage digestibility and by extension energy availability. Two herbages with similar true digestibility may differ greatly in how that value is achieved. They may have different concentrations of cell solubles and digestible fiber and rate of fiber digestion. Based on the model (Eq. (20)), strategies for increasing the true digestibility of herbage could include simply increasing the cell soluble concentration, increasing the concentration of digestible fiber, and/or increasing the rate at which the latter is digested. Focusing on improving any one of these parameters in isolation of the others would not necessarily lead to an improvement in true digestibility because whatever gains achieved in one could be lost from negative changes in the others. Using any or all of these strategies, the end result is a decrease in the concentration of indigestible fiber

(C_l), which in the end makes sense because it is the mathematical inverse of true digestibility (i.e., TD = 1000 − C_l). As it turns out, there is an additional benefit to decreasing indigestible fiber that can be described with a simple mathematical model.

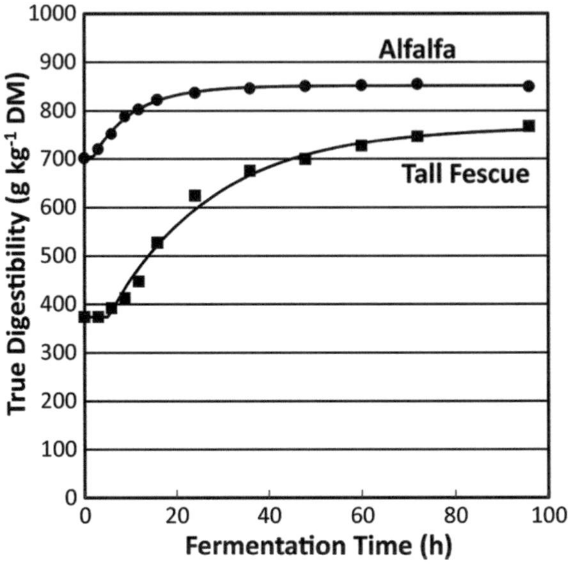

Figure 5. True digestion of forage dry matter during fermentation of alfalfa (●) and tall fescue (■). The values estimated by the line were calculated using Eq. (20). The point where each line intersects the ordinate represents the cell soluble concentration, which in Eq. (20) is considered to be completely digested. The overall picture that emerges is quite different from that where fiber digestion is considered in isolation. Data from Ref. [14].

9. Dry matter intake

The nutritional value of herbage depends largely on the amount of digestible energy that the animal derives from consuming it [27]. In this application, we present a model for predicting true digestibility that relates directly to the energy concentration available to support maintenance and production. How much energy the animal actually ingests, however, also depends on the amount of herbage consumed. Digestible energy intake is the product of dry matter intake and digestibility and is often limited for herbage diets.

There are many factors that influence the amount of herbage that is consumed by an animal. Some of these are related to the animal and its body size, plane of nutrition, and psychogenic factors that influence palatability [28]. There are chemostatic controls that regulate intake and tend to suppress it once the animal's demand for energy has been satisfied [29]. Intake of diets that are predominantly herbage, however, often are regulated by physical distention of the digestive tract. This latter mechanism is generally referred to as fill volume because it represents the quantity of undigested herbage than can be accommodated by the size of the digestive system.

The intake of indigestible fiber is often observed to be relatively constant across similar herbage diets that vary in digestibility suggesting a limit in the amount of indigestible material that an animal can consume [30]. As the digestibility of the diet increases for animals of similar size and nutritional status, the amount of dry matter that can be consumed also increases because there is less undigested material to retard its passage through the digestive system. Because of this, dry matter intake is often correlated to indigestible fiber concentration and a simple fill model can be used predict it:

$$DMI = \frac{F}{C_I} = \frac{F}{1 - TD} \tag{21}$$

where F is a fill constant and has the same units as DMI. It represents the intake capacity for indigestible fiber. Intake is often expressed as a percent of animal body weight (% BW), so in this case, F would represent the daily intake capacity for indigestible fiber expressed as a percentage of body weight. The concentration of indigestible fiber then should be expressed as a decimal proportion of forage dry matter. It is possible to linearize this equation so that F can be estimated from experimental data but is easier just to calculate the average indigestible fiber intake across a range of forages with varying DMI and indigestible fiber concentrations.

In a brief survey of the literature Moore et al. [31] found that growing beef steers consumed between 0.4 and 0.6% of their body weight of indigestible fiber when fed diets consisting of warm-season grasses. A graph showing predicted intake as a function of C_I using a fill constant of 0.5 is presented in **Figure 6**. Using this relationship, the estimated DMI would be 3.4% BW for alfalfa and 2.2% BW for tall fescue evaluated in Section 7. These are realistic estimates and could be reasonably accurate as long as the fill constant is similar for the class of animal consuming these diets.

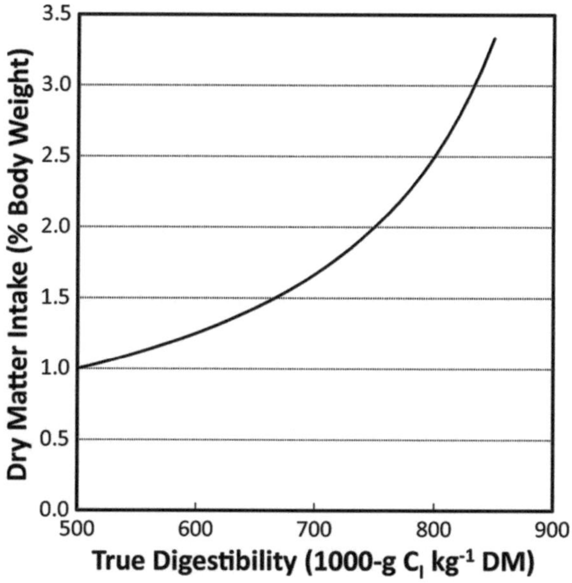

Figure 6. Predicted dry matter intake of warm-season grasses by growing beef steers using a fill constant of 0.5% of animal body weight (Eq. (21)). Adapted from Ref. [31].

It should be obvious that something as complex as DMI cannot be universally predicted using a simple model with one parameter. However, that is beside the point when using the model to develop strategies for improving forage quality. There should be no disagreement that for a given animal, there is a physical limit on how much indigestible dry matter they can consume. However, assuming that it is the same for all animals or even all animals within a specific class is probably unreasonable. This does not negate the utility of the concept for understanding how indigestible fiber affects DMI and nutritive value of herbage. The model is useful in that it demonstrates why modest improvements in true digestibility usually result in disproportionate increases in digestible energy intake [32].

10. Considerations

The value of using a simple model to describe biological responses is that it enables a better understanding of the response. It is one thing to say that observed values are different, another to say how they are different, and still yet another to say why they are different. Fitting a model to the response creates the possibility of accomplishing all three outcomes. It is important to realize, however, that the parameters of some models that fit a response cannot be easily interpreted. The coefficients from a quadratic equation used to fit the data from Section 5 would be difficult to interpret relative to any biological meaning or significance even though the model fits reasonably well ($r^2 = 0.81$ and 0.97 for alfalfa and tall fescue, respectively). Knowing that the digestion of fiber follows first-order kinetics is much more informative and the logical conclusions that can be made once this is accepted are quite useful.

The examples presented in this chapter demonstrate the utility of using simple mathematical models to explain nutritional aspects of herbivory. It should be understood that simple models cannot be expected to fully explain complex phenomena. There are too many factors involved in most biological systems to be able to do so. This does not mean that the models are not valid within the constraints they are used, but that they should not be generalized to other situations without validating their predictive performance in those situations.

Appendix

A1. Rumen degradable protein

Dataset

Bromegrass		Switchgrass	
CP	RDP	CP	RDP
188.9	148.1	131.1	68.1
144.5	106.0	118.9	61.4
144.5	101.4	106.3	50.3
124.2	80.2	113.3	52.1

http://dx.doi.org/10.5772/67342

Bromegrass		Switchgrass	
CP	RDP	CP	RDP
132.4	88.7	92.5	42.3
143.8	88.6	82.0	33.5
81.8	57.5	77.2	48.0
32.1	25.3	53.3	26.8
45.1	30.4	34.4	10.2
31.3	19.9	30.0	8.0
33.6	16.8	27.5	4.5
		18.3	2.0

CP = crude protein concentration (g kg^{-1} DM); RDP = rumen degradable protein (g kg^{-1} DM).

SAS Code

```
proc reg;
 by species;
 model RDP = CP;
run;
```

SAS Output *(abridged)*

```
----------------------------------- Species=Brome -----------------------------------

                                 Analysis of Variance
                                        Sum of           Mean
     Source                   DF       Squares         Square    F Value    Pr > F
     Model                     1         17755          17755     329.97    <.0001
     Error                     9     484.27891       53.80877
     Corrected Total          10         18239

                  Root MSE               7.33545    R-Square     0.9734
                  Dependent Mean        69.35455    Adj R-Sq     0.9705
                  Coeff Var             10.57673


                                 Parameter Estimates
                      Parameter      Standard
  Variable    DF       Estimate         Error   t Value   Pr > |t|     Type I SS    Type II SS
  Intercept    1       -4.70460       4.63830     -1.01     0.3369         52911      55.35812
  CP           1        0.73911       0.04069     18.17     <.0001         17755         17755

----------------------------------- Species=Switch -----------------------------------

                                 Analysis of Variance
                                        Sum of           Mean
     Source                   DF       Squares         Square    F Value    Pr > F
     Model                     1    5720.58349     5720.58349     238.45    <.0001
     Error                    10     239.90318       23.99032
     Corrected Total          11    5960.48667

                  Root MSE               4.89799    R-Square     0.9598
                  Dependent Mean        33.93333    Adj R-Sq     0.9557
                  Coeff Var             14.43416

                                 Parameter Estimates
                      Parameter      Standard
  Variable    DF       Estimate         Error   t Value   Pr > |t|     Type I SS    Type II SS

  Intercept    1       -8.25067       3.07601     -2.68     0.0230         13818     172.59967
  CP           1        0.57212       0.03705     15.44     <.0001    5720.58349    5720.58349
```

A2. Fiber digestion

Dataset

	Alfalfa				Tall fescue			
h/Rep	**1**	**2**	**3**	**4**	**1**	**2**	**3**	**4**
0	293.2[1]	296.7	297.1	306.2	642.3	617.5	639.4	604.9
3	276.6	291.3	274.4	276.6	637.2	615.1	632.4	603.0
6	237.2	254.9	242.7	257.7	623.3	605.0	608.7	594.9
9	208.1	218.9	205.4	214.3	603.8	593.9	581.3	571.3
12	194.7	195.8	190.6	207.4	555.1	561.3	551.5	544.3
16	177.1	180.2	168.9	182.6	461.2	489.4	471.8	467.1
24	169.9	170.6	155.4	156.8	394.8	366.7	373.0	367.2
36	151.2	146.3	152.7	165.5	307.5	326.1	345.2	320.1
48	146.9	144.6	152.2	154.0	299.9	307.7	316.4	277.9
60	145.3	142.9	147.6	154.4	260.0	285.1	281.3	264.5
72	142.8	141.5	143.3	153.4	241.2	264.6	249.8	257.9
96	148.3	147.8	151.4	154.3	219.0	243.1	238.3	228.6
120	137.4	136.8	141.5	150.2	200.9	236.9	207.7	201.3

[1]Values are neutral detergent fiber (NDF) concentration at h (g kg^{-1} DM).

A2.1. Determining time intervals to include in regression

SAS Code

```
proc glm;
by species;
class h;
model NDF = h / ss3;
means h / lsd hovtest;
run;
```

***SAS Output** (abridged)*

```
----------------------------------- species=Alfalfa -----------------------------------

          Levene's Test for Homogeneity of NDF Variance
          ANOVA of Squared Deviations from Group Means
                          Sum of         Mean
Source           DF      Squares       Square     F Value     Pr > F
h                11      15155.3       1377.8        1.10     0.3910
Error            36      45202.8       1255.6

                     t Tests (LSD) for NDF
              Alpha                                0.05
              Error Degrees of Freedom               36
              Error Mean Square                44.07785
              Critical Value of t               2.02809
              Least Significant Difference        9.521

Means with the same letter are not significantly different.
        t Grouping          Mean       N     h
                A        298.300       4     0
                B        279.725       4     3
                C        248.125       4     6
                D        211.675       4     9
                E        197.125       4    12
                F        177.200       4    16
                G        163.175       4    24
           H    G        153.925       4    36
           H             150.450       4    96
           H             149.425       4    48
           H             147.550       4    60
           H             145.250       4    72


----------------------------------- species=Fescue -----------------------------------

          Levene's Test for Homogeneity of NDF Variance
          ANOVA of Squared Deviations from Group Means
                          Sum of         Mean
Source           DF      Squares       Square     F Value     Pr > F
h                11       154556      14050.6        0.73     0.6997
Error            36       689475      19152.1

                   t Tests (LSD) for NDF
              Alpha                                0.05
              Error Degrees of Freedom               36
              Error Mean Square                180.9521
              Critical Value of t               2.02809
              Least Significant Difference       19.291


Means with the same letter are not significantly different.
      t Grouping          Mean       N     h
               A        626.025      4     0
               A        621.925      4     3
               A        607.975      4     6
               B        587.575      4     9
               C        553.050      4    12
               D        472.375      4    16
               E        375.425      4    24
               F        324.725      4    36
               G        300.475      4    48
               H        272.725      4    60
               I        253.375      4    72
               J        232.250      4    96
```

A2.2. Calculating rate constants using linear regression

SAS Code

Note that this procedure requires that time intervals where no digestion occurred have been deleted from the active data set, the indigestible fiber concentration ($[C_I]$) has been subtracted from the fiber concentration ($[C]_t$) remaining at each time point, and that this difference has been transformed by taking the natural logarithm. This quantity is included in the model statement as lnCD.

```
proc reg;

by species;

model lnCD = h / lackfit;

run;
```

The lackfit option in the model statement requests that the residual variance be partitioned into lack of fit and pure error in order to test if the model describes the response.

*SAS **Output**(abridged)*

```
------------------------------------ species=Alfalfa ------------------------------------

                              Analysis of Variance
                                     Sum of            Mean
      Source                  DF    Squares          Square    F Value    Pr > F

      Model                    1   14.17745        14.17745     220.39    <.0001
      Error                   22    1.41526         0.06433
        Lack of Fit            4    0.05980         0.01495       0.20    0.9359
        Pure Error            18    1.35546         0.07530
      Corrected Total         23   15.59271

                              Parameter Estimates

                                Parameter        Standard
           Variable      DF      Estimate           Error    t Value    Pr > |t|

           Intercept      1       5.20629         0.10175      51.17      <.0001
           h              1      -0.11145         0.00751     -14.85      <.0001

------------------------------------ species=Fescue ------------------------------------

                              Analysis of Variance
                                     Sum of            Mean
      Source                  DF    Squares          Square    F Value    Pr > F
      Model                    1   30.23112        30.23112     467.38    <.0001
      Error                   30    1.94045         0.06468
        Lack of Fit            6    0.35437         0.05906       0.89    0.5152
        Pure Error            24    1.58608         0.06609
      Corrected Total         31   32.17158

                              Parameter Estimates
                                Parameter        Standard
           Variable      DF      Estimate           Error    t Value    Pr > |t|
           Intercept      1       6.20922         0.08401      73.91      <.0001
           h              1      -0.04431         0.00205     -21.62      <.0001
```

Author details

Kenneth J. Moore*, Sotirios V. Archontoulis and Andrew W. Lenssen

*Address all correspondence to: kjmoore@iastate.edu

Iowa State University, Ames, IA, USA

References

[1] Archontoulis SV, Miguez FE. Nonlinear regression models and applications in agricultural research. Agronomy Journal. 2013; **107**:786-798. DOI: 10.2134/agronj2012.0506

[2] Petersen RG. Use and misuse of multiple comparison procedures. Agronomy Journal. 1977; **69**:205-208. DOI: 10.2134/agronj1977.00021962006900020003x

[3] Draper NR, Smith H. Applied Regression Analysis, 3rd edition. New York: Wiley; 1998. 736 p. ISBN: 9780471170822

[4] Pegelow J Jr, Taylor BB, Horrocks RD, Buxton DR, Marx DB, Wanjura DF. The Gompertz function as a model for cotton hypocotyl elongation. Agronomy Journal. 1977; **69**:875-878. DOI: 10.2134/agronj1977.00021962006900050038x

[5] Ware GO, Ohki K, Moon LC. The Mitscherlich plant growth model for determining critical nutrient deficiency levels. Agronomy Journal. 1982; **74**:88-91. DOI: 10.2134/agronj1982.00021962007400010024x

[6] Wraith JM, Or D. Nonlinear parameter estimation using spreadsheet software. Journal of Natural Resources and Life Science Education. 1998; **27**:13–19.

[7] Lucas HL Jr, Smart WWG Jr, Mart WW Jr, Cipolloni MA, Gross HD. Relations between digestibility and composition of feeds and foods. S-45 Report, North Carolina State College. 1961. As cited by [8].

[8] Van Soest PJ. Development of a comprehensive system of feed analyses and its application to forages. Journal of Animal Science. 1967; **26**:119-128. DOI: 10.2134/jas1967.261119x

[9] Van Soest PJ. The uniformity and nutritive availability of cellulose. Federation Proceedings. 1973; **32**:1804-1808.

[10] Van Soest PJ. Nutritional Ecology of the Ruminant, 2nd edition. Ithaca, NY: Comstock Publishing, Cornell University Press. 1994. 488 p. ISBN: 978-0801427725

[11] Moore KJ, Hatfield RD. Carbohydrates and forage Quality. In: Fahey GC Jr., et al. editors. Forage Quality, Evaluation, and Utilization. Madison, WI: American Society of Agronomy; 1994. p. 229-280. ISBN: 9780891181194

[12] Broderick, GA. Quantifying forage protein quality. In: Fahey GC Jr. et al. editors. Forage Quality, Evaluation and Utilization, Madison, WI: American Society of Agronomy; 1994. p. 200–228. ISBN: 9780891181194

[13] Mullahey JJ, Waller SS, Moore KJ, Moser LE, Klopfenstein TJ. In situ ruminal protein degradation of switchgrass and smooth bromegrass. Agronomy Journal. 1992; **84**:183-188. DOI: 10.2134/agronj1992.00021962008400020012x

[14] Moore KJ, Cherney JH. Digestion kinetics of sequentially extracted cell wall components of forages. Crop Science. 1986; **26**:1230-1234. DOI: 10.2135/cropsci1986.0011183X002600060032x

[15] Moore KJ, Jung H-JG. Lignin and fiber digestion. Journal of Range Management. 2001; **54**:420-430. DOI: 10.2307/4003113

[16] Smith LW, Goering HK, Waldo DR, Gordon CH. (1971) In vitro digestion rate of forage cell wall components. Journal of Dairy Science. 1971; **54**:71–76.

[17] Waldo DR, Smith LW, Cox EL. Model of cellulose disappearance from the rumen. Journal of Dairy Science. 1972; **55**:125-129.

[18] Cherney JH, Volenec JJ, Nyquist WE. Sequential fiber analysis of forage as influenced by sample weight. Crop Science. 1985; **25**:1113-1115. DOI: 10.2135/cropsci1985.0011183X002500060051x

[19] Phillips GR, Harris JM, Eyring EM. Treatment of replicate measurements in kinetic analysis. Analytical Chemistry. 1982; **54**:2053–2056.

[20] Zar JH. Biostatistical Analysis, 5th edition. Upper Saddle River, New Jersey: Prentice Hall; 2010. p. 944. ISBN: 9780131008465

[21] Varga GA. Factors which affect estimation of lag time in the rumen. In: International Symposium on Feed Intake by Beef Cattle; Oklahoma City, Oklahoma, 20-22 November 1986; 1986. p. 70-80.

[22] Mertens DR. Rate and extent of digestion. In: Dijkstra J, Forbes JM, France J, editors. Quantitative Aspects of Ruminant Digestion and Metabolism. Cambridge: CAB International; 2005. p. 13-47. ISBN: 9780851988313

[23] Moore TC. Biochemistry and Physiology of Plant Hormones. New York: Springer-Verlag; 1979. p. 274. ISBN: 978-1-4612-8193-1

[24] Ørskov ER, McDonald I. The estimation of protein degradability in the rumen from incubation measurements weighted according to rate of passage. Journal of Agricultural Science. 1979; **92**:499-503.

[25] Sleugh BB, Moore KJ, Brummer EC, Knapp AD, Russell JG, Gibson L. Forage nutritive value of various Amaranth species at different harvest dates. Crop Science. 2001; **41**:466-472. DOI: 10.2135/cropsci2001.412466x

[26] Moore KJ, Buxton DR. Fiber composition and digestion of warm-season grasses. In: Moore KJ, Anderson BA editors. Native Warm-Season Grasses: Research Trends and Issues. CSSA Special Publication, Madison, WI: Crop Science Society of America; 2000. p. 23-33. ISBN: 9780891185529

[27] Buxton DR, Mertens DR, Moore KJ. Forage quality for ruminants: plant and animal considerations. The Professional Animal Scientist. 1995; **11**:121-131. DOI: 10.15232/S1080-7446(15)32575-4

[28] Mertens DR. Predicting intake and digestibility using mathematical models of ruminal function. Journal of Animal Science. 1987; **64**:1548-58. DOI: 10.2134/jas1987.6451548x

[29] Moore JE. Forage crops. In: Hoveland CS editor. Crop Quality, Storage, and Utilization. Madison, WI: American Society of Agronomy; 1980. p. 61-91. ISBN: 9780891180357

[30] Moore KJ, Lemenager RP, Lechtenberg VL, Hendrix KS Risk JE. Digestion and utilization of ammoniated grass-legume silage. Journal of Animal Science. 1986; **62**:235-243. DOI: 10.2134/jas1986.621235x

[31] Moore KJ, Vogel KP, Pedersen JF. Improving the forage quality of warm-season perennial grasses. In: Proceedings of the American Forage and Grassland Conference; 1-4 April; Columbia. Missouri; 1991. p. 103-106.

[32] Moore KJ, Vogel KP, Hopkins AA, Pedersen JF, Moser JF. Improving the digestibility of warm-season perennial grasses. In: Proceedings of the XVII International Grassland Congress; 8–21 February; 1993. 19p. p. 447-448.

Chapter 5

Effects of Grain Processing with Focus on Grinding and Steam-Flaking on Dairy Cow Performance

Khalil Safaei and WenZhu Yang

Additional information is available at the end of the chapter

http://dx.doi.org/10.5772/67344

Abstract

Milk production and milk components are of prime economic importance for dairy farmers. Although milk production depends largely on numerous dietary nutrients, energy and protein are most critical. Feed grains containing starch such as corn, barley, wheat, and sorghum as a primary source of energy are commonly fed to beef and dairy cattle to improve meat or milk productions. Feed grain needs to be processed prior to feed cattle to increase accessibility of the endosperm by microbial population in the rumen and the host enzyme in the intestine. Grain processing is done by the application of various combinations of heat, moisture, time and mechanical actions. This article outlines the effect of grain processing method and degree of processing on rate and extent of grain digestion in the digestive tract of cattle, and consequently on lactation performance and cattle health. Methods of grinding, rolling and steam flaking are particularly discussed on their advantages and disadvantages. The optimal degree of processing can achieve a balance between maximizing the extent and controlling the rate of starch digestion in the rumen to maximize utilization and avoid digestive and metabolic disturbances. A recent developed precision processing technique has been highlighted and discussed as well.

Keywords: grain, processing, grinding, steam-flaking, dairy cow

1. Introduction

It is well known that the diet fed to dairy cows is an important lever by which milk yield and milk composition could be modified. Although milk production is affected by numerous dietary nutrients, energy and protein are most critical. Feed grains containing starch such as corn, wheat, barley, and sorghum, as a primary source of energy, are commonly fed to livestock to improve meat or milk productions [1]. Improving starch utilization may improve lactation performance in cows and reduce feed costs, especially when grain price is high [2]. Whole grain with an intact pericarp is largely or entirely resistant to digestion by ruminants because whole kernels are resistant to bacterial and host enzyme accessing to endosperm of grain kernel in the rumen and in the intestine, respectively [3, 4]. Therefore, cereal grains require processing to break the protective seed coat, and the fibrous hull in the case of barley and oats, and to improve grain digestibility [5]. Grain can be processed by the application of various combinations of heat, moisture, time, and mechanical action. Nonthermal processes (roller and hammer mill) and thermal processes (roasting, popping, micronizing, autoclaving, steam-flaking, steam pelleting, expanding, extruding, and toasting) could be used to manipulate rate of degradation and hence ruminal availability [6, 7]. The extent and rate of ruminal digestion of grain can be manipulated through processing [8]. The quality of processed grain can be affected by the application of various combinations of heat, moisture, time, and mechanical action [9]. These processing treatments alter kernel structure, thereby enhancing the release of starch granules from the protein matrix and disrupting their order during gelatinization; this, in turn, increases the accessibility of starch to microbes in the rumen and increases the susceptibility to enzyme activity [10]. Thus, grain processing can be a useful tool for optimizing lactating dairy cow production by synchronizing energy and protein to improve rumen microbial protein production. Steam-flaking is a more extensive processing system than dry- or steam-rolling [11]. Steam-flaking has been shown to increase the digestibility of starch by cattle fed corn- or sorghum grain-based diets over whole, ground, or dry-rolled grain [7]. In sorghum, steam-flaking has been found to increase ruminal starch digestion, as compared with dry-rolling or grinding [12]. Consistent increases in ruminal starch digestion have also been observed for steam-flaked corn, as compared with ground or dry-rolled corn [13]. Malcolm and Kiesling [14] reported that steam-flaking of barley tended to increase in situ rumen dry matter degradation. These experiments have shown that steam-flaking increases the amount of starch fermented in the rumen and, also, enhances the intestinal digestibility of starch that escape the rumen degradation. Few studies have investigated the effects of steam-flaked barley (SFB) on its digestibility in the rumen and in the total digestive tract of lactating dairy cows. Increasing the extent of grain processing enhances ruminal starch digestion of grain [8]. Additionally, grinding is a conventional processing method used in grain processing for feeding dairy cattle in some countries like Iran because of its low cost, but ground barley is often dusty, thereby potentially reducing feed intake. This response is usually influenced by either the extent of processing or flake density [13]. The objective of this chapter was to discuss the effects of grain processing methods with emphasizes on grinding and steam-flaking and on grain digestibility and lactation performance of dairy cows.

2. Starch in cereal grains

Cereal grains are rich in starch ranging from 40% in oats up to 80% of dry matter in rice, with the variation in starch content dependent on variety, climatic conditions, and agronomic practices [15]. Starch is synthesized into a form of rough spherical granules and, within each feed grain starch granule, multiple concentric semicrystalline and amorphous shells are present [1].

Feed grain starch consists of amylose and amylopectin, as two major components [1, 5], which are present in different proportions in the starch granule of feed grains [1]. Cereal starches are typically composed of 25–28% amylose and 72–75% amylopectin. There are also waxy cultivars with very high amylopectin concentration (up to 100%) and high amylase (up to 70% amylose) cultivars [5]. Starch of waxy barley varieties may be less digestible than normal barley due to the differences in chemical composition and starch structure. Starch granules isolated from different feed grains and other starch sources reveal characteristic morphologies, varying in shape [16], in molecular size [17], and in specific surface structure and porosity [18]. Also, it has been recognized that cultivars or varieties in feed grains vary in granule size distribution, suggesting a significant genetic control [19]. Other components of feed grains, including lipids, proteins, and minerals, are also associated with starch granules. Lipids and proteins are the most abundant nonstarch components in feed grains that may affect physical state and enzyme susceptibility of starch in livestock [20].

3. Role of grain processing in starch digestion

Whole grains with an intact pericarp are largely or entirely resistant to digestion by ruminants because whole kernels are resistant to bacterial attachment in the rumen [3, 4]. Unlike sheep, adult cattles have a limited capacity to masticate cereal grains; hence, it is essential to break the pericarp of the seed either through chemical or physical treatments [21]. In feed grains, germ and endosperm are surrounded by the pericarp, which is largely resistant to microbial attachment [22]. Starch granules from corn grain are surrounded by a protein matrix in the endosperm [23], which influences digestion by microorganisms [24]. Feed grain endosperm encapsulated by protein matrix acts as a physical barrier to protect from enzymatic hydrolysis [25]. It has been shown that this protein matrix by blocking the absorption sites or by influencing enzyme binding may reduce the surface availability of starch to host enzyme and ruminal bacteria [25]. In addition, the results of several studies have shown that hydrophobic properties of grain protein matrix, associated with type and location of proteins, could be responsible for the differences in starch digestion between rapidly digested grains such as wheat, barley, and rice, and slowly digested grains such as maize and sorghum [2, 24]. Processing is necessary for feed grains to break the protective seed coat [3, 5] especially for the grains such as maize and sorghum [11], and the fibrous hull in case of barley and oats, to improve their digestibility in the digestive tract of animals [5]. Feed grain processing by grinding, rolling, pelleting, steam-rolling, or steam-flaking breaks down the recalcitrant barriers such as the hull, pericarp, and protein matrix and allows microbial and host enzyme accessing to the starch within endosperm cells.

4. Impact of grain processing

Grain processing methods could be divided into two groups: (1) nonthermal processes such as roller and hammer mill and (2) thermal processes, which include dry processing (roasting, popping, and micronizing) and wet processing (autoclaving, steam-flaking, steam-pelleting, expanding, extruding, and toasting) [26]. Heat processing, however, has been associated with increased efficiency of fermentative utilization by altering the protein matrix of the endosperm and the starch structure, thus allowing a better utilization by microbial enzymatic digestion [21]. Increasing the starch availability is the primary goal of grain processing. In addition, processing may destroy mycotoxins and improve mixing characteristics to improve bunk management and thereby enhance animal performance [27]. The processes reduce the particle size of the grain, increasing the surface area available for microbial attachment and colonization; combined, these actions increase the rate and extent of starch digestion [22]. Starch gelatinization, a process in which disaggregated amylose and amylopectin chains in a gelatinized starch paste reassociate to form more ordered structures, and dextrination, formation of dextrins (fragments of amylose and amylopectin molecules formed by heating dry starch in the presence of some moisture, acids, or salts) during processing of grains [28], improve accessibility of enzymes to the starch granules. It, in turn, may shift the site of digestion of protein and starch from the rumen to the intestine [6, 7], and consequently, it results in an improved supply of amino acids and glucose to animal metabolism [29]. Elevated glucose absorption represents one mechanism by which increased intestinal starch digestion might increase milk yield [30]. Increased net energy density of cereals is also beneficial because high-yield dairy cows often are unable to consume sufficient energy during early lactation to meet the requirements.

5. Grain starch digestion in digestive tract of dairy cows

Ruminants do not produce appreciable quantities of salivary amylase, and therefore, starch digestion in ruminants is initiated after ingestion and mastication of feed [1]. Dairy cattle consume large amounts of starch (20–40% of diet DM) as a way to increase energy consumption in support of high milk production [31]. The optimum starch content of diets for lactating dairy cows is not well defined, but 24–26% starch (dry matter basis) has been suggested [32]. Total tract starch digestibility of dairy cows is highly variable and has a range of 70–100% [33]. Quantitatively, most of the starch digestion in ruminants [2] occurs in the rumen (55–85%) and the small intestine (15–25%) with indigestible starch fractions ranging from 0 to 20% of starch intake [2, 34]. The rate and extent of starch digestion in the rumen are determined by interrelations among several factors, including level of feed intake, source of dietary starch, diet composition, grain processing, chemical alterations (fermentation, gelatinization), grain passage rate, and degree of adaptation of rumen bacteria to dietary starch [35]. Digestion of starch in the rumen is facilitated symbiotically by ruminal bacteria with amylolytic capacity [35]. Attachment of bacteria to starch containing feed particles is a key step in bacterial fermentation of starch in the rumen [1]. Although ruminal protozoa play a role in ruminal starch digestion through ingesting and digesting starch granules, eliminating protozoa by rumen defaunation can actually increase rate and extent of ruminal starch digestion [36], which suggests that bacterial fermentation alone is sufficient in starch digestion in the rumen. In general, ruminal starch digestion can vary substantially, ranging from 40 to 80% in lactating dairy cows [2, 33]. Compared to beef cattle, the extent of ruminal starch digestion from maize grain is often lower for lactating dairy cows [34]. The high levels of feed intake, rapid rates of passage and minimal mastication of grain particles by lactating dairy cows are thought to be responsible for the lower ruminal starch digestion [33, 34]. Approximately 50–90% of the postruminal flow of starch is digested in the small intestine [1, 3] by enzymatic hydrolysis of starch that provides energy in the form of glucose to the animal. The passage rate through the small intestine, surface exposure, and grain endosperm properties can alter the output of enzymatic starch hydrolysis in the small intestine [37]. In the large intestine, around 30–60% of the starch escaping ruminal digestion and enzymatic hydrolysis can be digested via hind gut bacterial fermentation [6] that is symbiotically providing energy in the form of volatile fatty acid.

6. Factors affecting grain starch digestion

The kinetics of starch digestion in livestock animals depend largely on two major factors: (1) the inherent starch architecture and related physicochemical properties and (2) the degree of processing feed grains prior to feeding that vary with methods used such as rolling, pelleting, flaking, extrusion, and expander processing as well as the processing condition setting up of each method [1]. Starches from various sources show different responses to the heat processing conditions. The starch structure can be modified by the application of such processing, thus some native physicochemical starch properties including starch granule morphology, crystallinity, the amylose content, and the type of endosperm are potentially affected [38]. These native physicochemical properties have been recognized to influence the starch digestion of grains [38]. Feed grains fed to livestock are commonly processed through a grinder or a roller prior to feeding, and mechanical processing could be considered as one of the primary and less expensive methods to influence starch digestibility. In commercial sector, various processing equipments with variable screen sizes or mill settings are available to create wide distributions of grain particle sizes and to meet different requirements of animal feeding [39]. The effect of particle size distribution on starch digestion is primarily associated with the available surface area for microbial access and enzymatic hydrolysis [40]. The particle size of grains after processing, even if not directly related to native physiochemical starch properties, also plays a role in determining the relative efficiency of starch digestion within grains [41]. Blasel et al. [41] reported a reduction in the degree of starch access by α-amylase of about 27 g/kg of starch for each 100 mm increase in particle size in ground maize grain. Also, it is reported a reduction in the rate of starch digestion of grounded feed grains through 3.0 mm compared with 0.8 mm opening screen after either 60 or 240 min of in vitro incubation [42]. In addition, it has been observed that there is an inverse square relationship between the enzymatic hydrolysis rates of pancreatic α-amylase and the increasing particle size in milled barley and sorghum grains [43].

6.1. Method of processing

The method of processing grain will depend on the type of grain and the specific feeding and management conditions. Small kernel grain, such as barley and wheat, is usually processed by dry-rolling, steam-rolling, or temper-rolling, while corn grain is often steam-flaked in cattle feeding. Grain can also be treated with chemicals or enzymes to alter the rate and extent of nutrient degradation [26]. Mechanical processing by rolling of the grain cracks or crushes the fibrous hull and pericarp to enable access of rumen microorganisms and enzymes to the internal endosperm and increases the surface area for attachment [22]. The addition of moisture in steam-rolling and temper-rolling may be advantageous over dry-rolling when the original grain is very dry (<10% moisture), of variable kernel size, rolled more extensively to maximize utilization, and fed with limited amount of forage. Steam-processed grain is exposed to steam at either atmospheric, low, or high pressure to increase grain moisture and temperature before the grain is rolled. Steam-flaking of grain is a more extensive process with longer steam conditioning times and thinner rolled flakes than steam-rolling. The combination of moisture, heat, and rolling causes gelatinization of the starch granules, i.e., swelling of granules as water is absorbed, disruption of the crystalline structure, dissolution of polysaccharides, and diffusion from ruptured granules. Steam-flaking of barley grain increased the in vitro amyloglucosidase, catalyzed release of glucose and the rate of starch degradation compared to dry-rolling of the grain [44]. However, the beneficial effects of starch gelatinization for barley may be less than for corn or sorghum grain because barley starch once it becomes accessible readily degraded by ruminal microorganisms [44].

6.2. Degree of processing

The optimal degree of grain processing is aimed at achieving a balance between maximizing the extent and controlling the rate of starch digestion in the rumen to increase utilization and avoid digestive and metabolic disturbances [5]. The degree of grain processing, also called the extent of processing, can significantly affect rate and extent of grain digestibility in the rumen or in the total digestive tract, thus affect feeding value and cattle performance as well as cattle health when the high-grain is fed. Overprocessed grain increases proportion of fine particles, reduces palatability, increases rate of grain digestion in the rumen, hence increases the incidence of digestive disorders such as rumen acidosis, bloat, laminitis, and liver abscesses [5]; whereas underprocessed grain reduces the starch availability for fermentation by rumen microbes [45]. The degree of grain processing has been widely used by cattle nutritionist to manipulate rate and extent of grain digestion in the rumen. Processing index was developed by our laboratory to quantify the degree of processing, and it is currently applied as a routine method by feed mill in grain processing for cattle feeding [4]. The processing index refers to the volume weight (g/L) of barley after processing expressed as a percentage of its volume weight before processing [4]. This index reflects the fact that the more extensively grain is processed (i.e., the higher the degree of processing), the finer the particle will be, hence, the lower the volume weight will be, and consequently, the lower the processing index. Take barley grain as an example; the disadvantages of under- or overprocessing grain on the milk production of dairy cattle were described in a study by Yang et al. [4]; barley grain was steam-rolled to four degrees with processing index of 81, 72, 64, and 55%, and milk production increased quadratically with degree of processing resulting, respectively, in increases of 0, 9.8, 20.3, and 13.3% in milk yield. The reduced milk yield in dairy cows fed coarsely rolled barley grain compared to those fed more extensively processed barley was due to the combined effects of lower feed intake, lower nutrient digestibility, and a slower particle outflow rate from the rumen. We concluded that the optimum extent of barley processing for dairy cows fed diets supplying adequate fiber was a processing index of 64% to maximize milk yield without negatively affecting milk fat percentage.

6.3. Precision processing

Kernel uniformity of grains can vary considerably depending on variety, growing conditions, disease, etc., in particular, grains with lower and higher volume weights are often blended commercially to achieve an intermediate volume weight that increases the market value of the grain [5], but it also increases the variability of kernel size. The poor kernel uniformity can contribute to inconsistent processing especially when the dry-rolling technique is used. We have developed the precision processing method [45–47] that refers to separate fractions of uniform kernel size and then each fraction is processed with an optimum roller setting specific to the kernel size to ensure that all kernels are cracked [46]. For barley grain of variable kernel size and shape, precision processing enables greater control over the extent of processing compared with processing with the conventional one roller setting, particularly when the grain is dry-rolled. Precision processing increased digestibility of crude protein and acid detergent fiber; as a result, organic matter digestibility increased as compared with the conventional processing in a study using beef cattle fed high-grain diet [46]. The barley that was sieved and rolled with settings based on kernel size had increased starch degradability up to 24 h and increased overall rate and effective degradability of starch in the rumen compared to the barley processed as a single roller setting [45]. In a lactation study, dry matter intake, ruminal fermentation, nutrient digestibility, and milk production were not affected by precision processing compared with conventional processing [47]. The differences in the results between the beef and dairy studies are likely due to the differences in the amount of barley grain (67 versus 40%) and forage (10 versus 40%) in the diets.

7. Effects of grain processing on dairy cow performance

7.1. Steam-flaking versus rolling or grinding

Compared with grinding, rolling processing generally reduces starch digestibility in the small intestine and in the total digestive tract of cattle [26]. Larsen et al. [48] in a review reported that reduction in ruminal starch digestion of grains was generally not compensated for by increasing the small intestinal starch digestion but was associated with increases of starch digestion in hindgut. Santos et al. [49] reported that dairy cows fed steam-flaked corn yielded daily 1.5 kg more milk than did cows fed rolled grain, also the yield of milk protein and percentages of lactose and solids that are nonfat increased compared with rolling processing. Similarly, compared with rolling, feeding steam-flaked corn and sorghum increased yields of milk, milk protein, and fat, and protein percentage of milk, which were likely due to the increased total tract digestibility of dry matter, crude protein, starch, and neutral detergent fiber [50]. It suggests that steam-flaking versus rolling improved the feeding value of corn for lactating cows by improving diet acceptability and digestibility of organic matter and increased the estimated net energy for lactation of corn. López-Soto et al. [51] indicated that compared to dry-rolled barley, steam-flaked barley increased ruminal digestion of organic matter and starch of diet, but decreased dry matter intake of lactating dairy cows. Our in situ experiment [52] showed that ground barley versus steam-flaked barley increased organic matter disappearance (**Figure 1**) and increased the washable fraction (28.3 versus 21.4%), rate of potentially degradable fraction (0.10 versus 0.05 h^{-1}), and effective degradability (60.6 versus 47.6%) of organic matter. In an in vivo study [53] using lactating dairy cows, we observed that the grinding versus steam-flaking barley grain did not affect dry matter intake (23.6 kg/day), digestibility of dry matter in the total digestive tract (71.0%), milk yield (43.4 kg/day), milk components, rumen pH, and molar proportions of acetate, propionate, and butyrate. The lack of difference between grinding and flaking could be due to low grain inclusion rate and minimal difference in the particle size as barley was coarsely ground. In contrast, Eastridge et al. [54] reported that the finely ground corn decreased milk urea nitrogen content compared with coarsely ground corn, suggesting that finely ground corn provided more fermentable starch in the rumen, thus possibly improved bacterial capture of the nitrogen. However, the results on dry matter intake, milk production of lactating dairy cows by feeding steam-flaked corn versus ground corn are inconsistent [55–57]. In another study using lactating dairy cows, we found that grinding versus steam-flaking of barley affected dry matter intake, organic matter digestibility in the total digestive tract, and feed efficiency. The discrepancy among the studies could be due to differences in grain variety, particle size distribution of processed grain, inclusion rate of grain in the diets, which could affect the grain digestibility in the rumen and in the intestine, hence, impact feed intake and milk production of dairy cows (**Figure 2**).

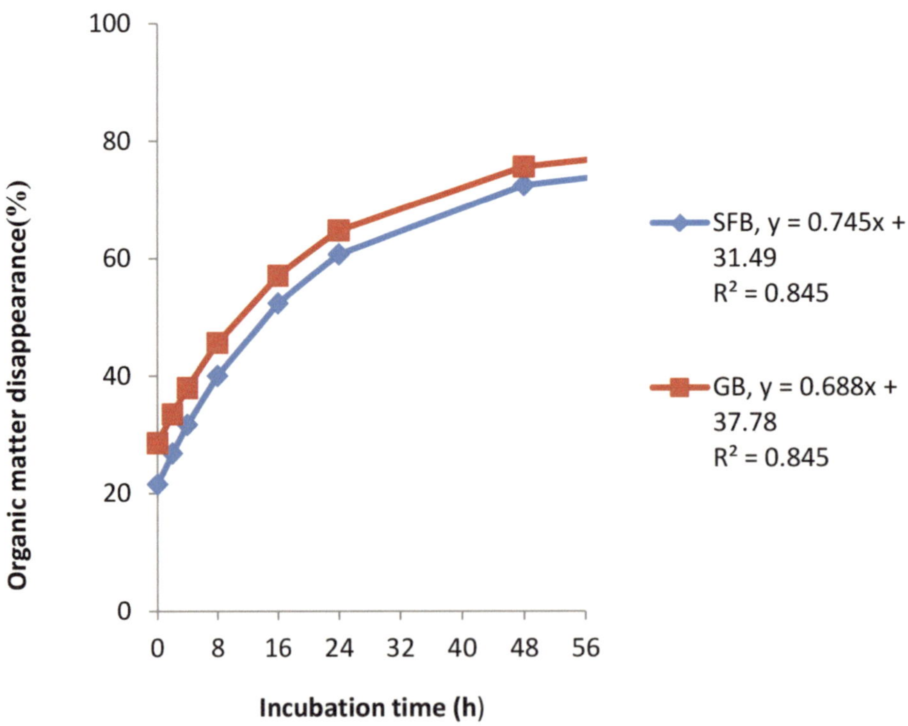

Figure 1. Organic matter disappearance (%) of ground barley (GB) compared to steam-flaked barley (SFB).

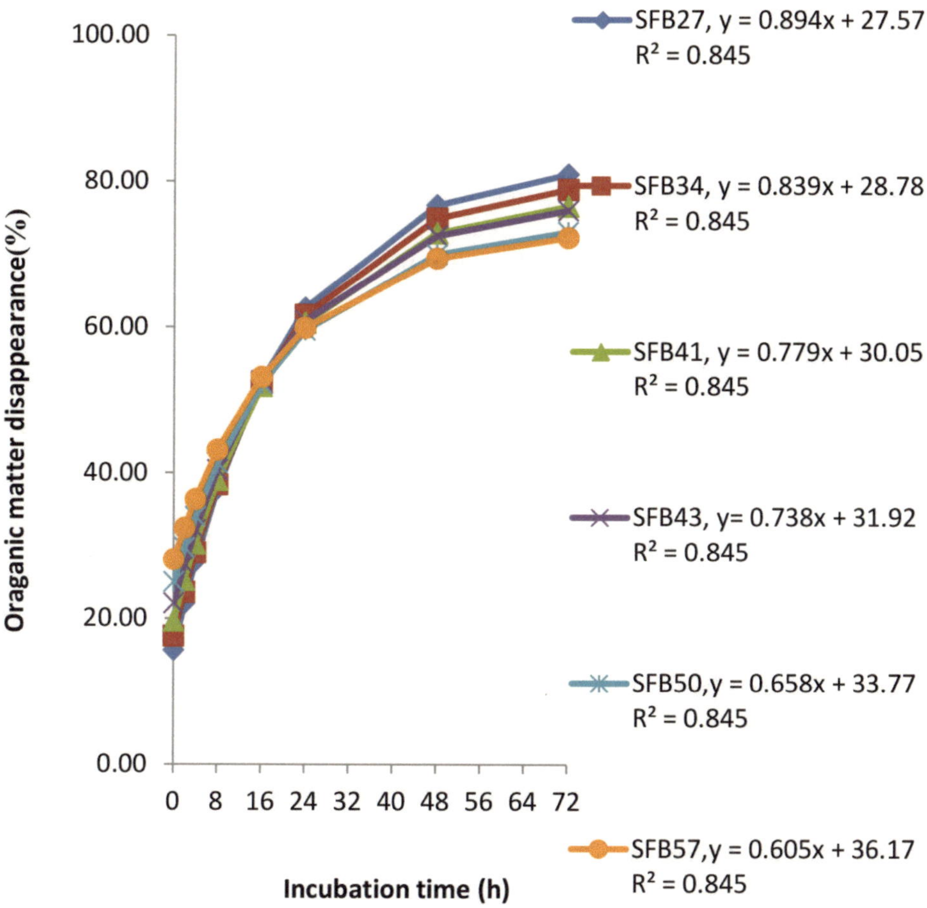

Figure 2. Organic matter disappearance (%) of steam-flaked barley (SFB) with various processing indexes (%).

7.2. Effect of flake density

Flake density of grains affects grain digestibility and cow performance. It was suggested that optimal flake density was between 0.32 and 0.39 kg/L [58]. Zinn and Barajas [59] conducted an experiment to evaluate the influence of flake density on feeding value of a barley-corn blend fed to feedlot cattle, and observed an increase of ruminal digestibility of organic matter, starch, and protein, but a decrease of ruminal nitrogen efficiency with decreasing flake density. We found that decreasing the density of steam flaked barley from 0.30 to 0.26 kg/L increased ruminal digestion of starch and ruminal propionate, but decreased dry matter intake and ruminal protein degradation. These results are expected since decreasing the density of processed barley indicated increased degree of processing, hence increased rumen starch digestibility (**Figure 2**), and consequently, decreased rumen pH (potential rumen acidosis), and decreased dry matter intake. In comparing with corn-based diet, dairy cows fed barley-based diets showed greater dietary energy due to improved ruminal microbial efficiency, greater total tract organic matter digestion, and lower ruminal acetate and methane production. However, dairy cows that were fed barley-based diets had lower ruminal pH, which was exacerbated as flake density decreased.

Results showed 8% improvement in energy value of barley with steam-flaking [51]. Flaking barley too thinly would depress feed intake because of increase in rumen digestion and reduction in rumen pH [51]. Our study also showed that decreasing the density of steam-flaked barley from 390 to 340 to 290 g/L tended to linearly decrease dry matter intake, total solids percentage of milk, and linearly decreased milk urea nitrogen. Finely ground corn increased ruminal propionate concentration and decreased ruminal pH and acetate to propionate ratio, suggesting an increase of grain digestion in the rumen. Increasing density of steam-flaked corn increased total tract digestion of organic matter, neutral detergent fiber, starch, and digestible energy content of diet and increased milk efficiency [60]. It suggests that barley and corn should be processed at different density when it is steam flaked to maximize its digestibility, while to minimize its rumen health problem since barley is more rapidly digestible than corn in the rumen.

8. Conclusion and implication

Improvement and optimization of grain digestion in the rumen and the intestine is an important research focus in cattle nutrition and feeding. Generally, site, extent, and rate of grain starch digestion in the digestive tracts of cattle are influenced by intrinsic and external factors that can be interrelated, hence they are not easily defined. Particle size reduction, starch gelatinization, retrogradation, and dextrination due to grain processing may shift the site of starch digestion from the rumen to the intestine, and thus it results in an improved supply of amino acids and glucose to animal. Steam-flaking grain increases the starch digested both in the rumen and in the intestine; this, in turn, increases the available energy for milk production. Overall, both ground and steam-flaked grains could be fed dairy cows depending on the level of grain in diet, dietary composition, and economic cost of grain processing.

Feeding values differ among grain sources and processing methods used. Although the net energy value of grain usually is increased by more extensive flaking, regardless of grain source or processing, dairy cows can produce milk at similar rates, probably due to reduction in dry matter intake and lower passage rate of flaked grains compared with ground grains. This may support the concept that chemostatic factors generally control intake of low forage diets. For a specific grain and processing method, roughage source and moisture may markedly influence rate of production and net energy value of the grain, probably due to bunk management, diet acceptability, chewing activity, and site and extent of starch digestion.

Author details

Khalil Safaei[1] and WenZhu Yang[2*]

*Address all correspondence to: wenzhu.yang@agr.gc.ca

1 Department of Animal Sciences, College of Agriculture, Isfahan University of Technology, Isfahan, Iran

2 Agriculture and Agri-Food Canada, Lethbridge Research and Development Centre, Lethbridge, Alberta, Canada

References

[1] Giuberti G, Gallo A, Masoero F, Ferraretto LF, Hoffman PC, Shaver RD. Factors affecting starch utilization in large animal food production system: A review. Starch Stärke. 2014;66:72–90.

[2] Ferraretto LF, Crump PM, Shaver RD. Effect of cereal grain type and corn grain harvesting and processing methods on intake, digestion, and milk production by dairy cows through a meta-analysis. J. Dairy Sci. 2013;96:533–550.

[3] Beauchemin KA, McAllister TA, Dong Y, Farr BI, Cheng KJ. Effects of mastication on digestion of whole cereal grains by cattle. J. Anim. Sci. 1994;72:236–246.

[4] Yang WZ, Beauchemin KA, Rode LM. Effects of barley grain processing on extent of digestion and milk production of lactating cows. J. Dairy Sci. 2000;83:554–568.

[5] Koenig, KM, Beauchemin KA, Yang WZ. Processing feed grains: Factors affecting the effectiveness of grain processing for beef and dairy cattle production. Proceedings of the 34th Western Nutrition Conference-Processing, Performance and Profit, Saskatoon, SK, Canada, 2013.

[6] Owens FN, Zinn RA, Kim YK. Limits to starch digestion in the ruminant small intestine. J. Anim. Sci. 1986;63:1634–1648.

[7] Theurer CB. Grain processing effects on starch utilisation by ruminants. J. Anim. Sci. 1986;63:1649–1662.

[8] Koenig KM, Beauchemin KA, Rode LM. Effect of grain processing and silage on microbial protein synthesis and nutrient digestibility in beef cattle fed barley-based diets. J. Anim. Sci. 2003;81:1057–1067.

[9] Kokić BM, Lević JD, Chrenková M, Formelová Z, Poláčiková M, Rajský M, Jovanović RD. Influence of thermal treatments on starch gelatinization and in vitro organic matter digestibility of corn. Food and Feed Res. 2013;40:93–99.

[10] Hoover R, Vasanthan T. Effect of heat-moisture treatment on the structure and physicochemical properties of cereal, legume, and tuber starches. Carb. Res. 1994;252:33–53

[11] Theurer CB, Huber JT, Delgado-Elorduy A, Wanderley R. Invited review: Summary of steam-flaking corn or sorghum grain for lactating dairy cows. J. Dairy Sci., 1999;82: 1950–1959.

[12] Hinman DD, Johnson RR. Influence of processing methods on digestion of sorghum starch in high concentrate beef cattle rations. J. Anim. Sci. 1974;39:417–422.

[13] Zinn RA. 1990. Influence of steaming time on site of digestion of flaked corn in steers. J. Anim. Sci. 1990;68:776–781.

[14] Malcolm KJ, Kiesling HE. Dry matter disappearance and gelatinization of grains as influenced by processing and conditioning. Anim. Feed Sci. Technol. 1993;40:321–330.

[15] Giuberti G, Gallo A, Cerioli C, Masoero F. In vitro starch digestion and predicted glycemic index of cereal grains commonly utilized in pig nutrition. Anim. Feed. Sci. Technol. 2012;174:163–173.

[16] Tester RF, Karkalas J. Swelling and gelatinization of oat starches. Cereal Chem. 1996;73: 271–277.

[17] Pérez S, Bertoft E. The molecular structures of starch components and their contribution to the architecture of starch granules: A comprehensive review. Starch Stärke. 2010;62:389–420.

[18] Singh J, Dartois A, Kaur L. Starch digestibility in food matrix: A review. Trends Food Sci. Technol. 2010;21:168–180.

[19] Peterson DG, Fulcher RG. Variation in Minnesota HRS wheats: Starch granule size distribution. Food Res. Int. 2001;34:357–363.

[20] Baldwin PM. Starch granule-associated proteins and polypeptides: A review. Starch Stärke. 2001;53:475–503.

[21] Alvarado CG, Anrique RG, Navarrete SQ. Effect of including extruded, rolled or ground corn in dairy cow diets based on direct cut grass silage. Chil. J. Agric. Res. 2009;69:356–365.

[22] McAllister TA, Bae HD, Jones GA, Cheng KJ. Microbial attachment and feed digestion in the rumen. J. Anim. Sci. 1994;72:3004–3018.

[23] Kotarski SF, Waniska RD, Thurn KK. Starch hydrolysis by the ruminal microflora. J. Nutr. 1992;122:178–190.

[24] McAllister TA, Phillippe RC, Rode LM, Cheng KJ. Effect of the protein matrix on the digestion of cereal grains by ruminal microorganisms. J. Anim. Sci. 1993;71:205–212.

[25] Drew MD, Schafer TC, Zijlstra RT. Glycemic index of starch affects nitrogen retention in grower pigs. J. Anim. Sci. 2012;90:1233–1241.

[26] Dehghan-Banadaky, M., Corbett, R., Oba, M. 2007. Effect of barley grain processing on productivity of cattle. Anim. Feed. Sci. Technol. 2007;137:1–24.

[27] Owens FN, Secrist DS, Hill WJ, Gill DR. The effect of grain source and grain processing on performance of feedlot cattle: A review. J. Anim. Sci. 1997;75:868–879.

[28] Rooney LW, Pflugfelder RL. Factors affecting starch digestibility with special emphasis on sorghum and corn. J. Anim. Sci. 1986;63:1607–1623.

[29] Nocek JE, Tamminga S. Site of digestion of starch in the gastrointestinal tract of dairy cows and its effect on milk yield and composition. J. Dairy Sci. 1991;74:3598–3629.

[30] Sutton JD. Digestion and absorption of energy substrates in the lactating cow. J. Dairy Sci. 1985;68:3376–3393.

[31] Patton RA, Patton JR, Boucher SE. Defining ruminal and total-tract starch degradation for adult dairy cattle using in vivo data. J. Dairy Sci. 2012;95:765–782.

[32] Staples CR, Feeding dairy cows when corn prices are high. Proceedings 44th Florida Dairy Production Conference, Gainesville, FL, USA, May 1, 2007, pp. 7–22.

[33] Firkins JL, Eastridge ML, St-Pierre NR, Noftsger SM. Effects of grain variability and processing on starch utilization by lactating dairy cattle. J. Anim. Sci. 2001;79:218–238.

[34] Owens FN. Corn grain processing and digestion. 66th Minnesota Nutrition Conference, St. Paul, MN, Sept. 2005; pp. 20–21

[35] Huntington GB. Starch utilization by ruminants: From basics to the bunk. J. Anim. Sci. 1997;75:852–867.

[36] Mendoza GD, Britton RA, Stock RA. Influence of ruminal protozoa on site and extent of starch digestion and ruminal fermentation. J. Anim. Sci. 1993;71:1572–1578.

[37] Van Kempen TA, Pujol TG, Tibble S, Belfagon S, in: Wiseman, J, Varley MA, McOrist S, Kemp B. (Eds.), Paradigms in Pig Science, Nottingham Univ. Press, Nottingham, UK. 2007; pp. 515–526.

[38] Zinn RA, Owens FN, Ware RA. Flaking corn: Processing mechanics, quality standards, and impacts on energy availability and performance of feedlot cattle. J. Anim. Sci. 2002;80:1145–1156.

[39] Remond D, Cabrera-Estrada JI, Champion M, Chauveau B, Coudure R, Poncet C. Effect of corn particle size on site and extent of starch digestion in lactating dairy cows. J. Dairy Sci. 2004;87:1389–1399.

[40] Al-Rabadi GJ, Torley PJ, Williams BA, Bryden WL, Gidley MJ. Particle size heterogeneity in milled barley and sorghum grains: Effects on physico-chemical properties and starch digestibility. J. Cereal Sci. 2012;56:396–403.

[41] Blasel HM, Hoffman PC, Shaver RD. Degree of starch access: An enzymatic method to determine starch degradation potential of corn grain and corn silage. Anim. Feed. Sci. Technol. 2006;128:96–107.

[42] Anguita M, Gasa J, Martín-Orúe SM, Pérez JF. Study of the effect of technological processes on starch hydrolysis, non-starch polysaccharides solubilization and physico-chemical properties of different ingredients using a two-step in vitro system. Anim. Feed. Sci. Technol. 2006;129:99–115.

[43] Al-Rabadi GJS, Gilbert RG, Gidley MJ. Effect of particle size on kinetics of starch digestion in milled barley and sorghum grains by porcine alpha-amylase. J. Cereal Sci. 2009;50:198–204.

[44] Zinn RA. Influence of processing on the comparative feeding value of barley for feedlot cattle. J. Anim. Sci. 1993;71:3–10.

[45] Ahmad M, Gibb DJ, McAllister TA, Yang WZ, Helm J, Zijlstra RT, Oba M. Adjusting roller settings based on kernel size increased ruminal starch digestibility of dry-rolled barley grain in cattle. Can. J. Anim. Sci. 2010;90:275–278.

[46] Yang WZ, Oba M, McAllister TA. Quality and precision processing of barley grain affected intake and digestibility of dry matter in feedlot steers. Can. J. Anim. Sci. 2013;93:251–260.

[47] Schlau N, Duineveld L, Yang WZ, McAllister TA, Oba M. Precision processing barley grain did not affect productivity of lactating dairy cows. Can. J. Anim. Sci. 2013;93:261–268.

[48] Larsen M, Lund P, Weisbjerg MR, Hvelplund T. Digestion site of starch from cereals and legumes in lactating dairy cows. Anim. Feed Sci. Technol. 2009;153:236–248.

[49] Santos JEP, Huber JT, Theurer CB, Nussio LG, Tarazon M, Santos FAP. Response of lactating dairy cows to steam-flaked sorghum, steam-flaked corn, or steam-rolled corn and protein sources of differing degradability. J. Dairy Sci. 1999;82:728–737.

[50] Chen KH, Huber JT, Theurer CB, Swingle RS, Simas SC, Chan ZWu, Sullivan JL. Effect of steam flaking of corn and sorghum grains on performance of lactating cows. J. Dairy Sci. 1994;77:1038–1043.

[51] López-Soto MA, Barreras A, Calderón-Cortés JF, Plascencia A, Urías-Estrada JD, Aguilar-Hernández JA, Sánchez-Mendoza B, Montelongo-Terriquez A, Bermúdez-Hurtado RM, Estrada-Angulo A, Zinn RA. Influence of processing of barley grain on characteristics of digestion, ruminal fermentation and digestible energy of diet in lactating cows. Iranian J. Appl. Anim. Sci. 2014;4:477–484.

[52] Safaei Kh, Ghorbani GR, Alikhani M, Sadeghi Sefidmazgi A, Yang WZ, Mohammadi F. Effects of processing method, steaming duration and roller setting distance on ruminal degradability of barley grain. J. Rum. Res. 2016;3:103–126.

[53] Safaei Kh, Ghorbani GR, Alikhani M, Sadeghi Sefidmazgi A, Yang WZ. Response of lactating dairy cows to degree of steam-flaked barley grain in low-forage diets. J. Anim. Physio. Anim. Nutr. 2016; doi: 10.1111/jpn.12565 [In Press].

[54] Eastridge ML, Lefeld AH, Eilenfeld AM, Gott PN, Bowen WS, Firkins JL. Corn grain and liquid feed as nonfiber carbohydrate sources in diets for lactating dairy cows. J. Dairy Sci. 2011;94:3045–3053.

[55] Cooke KM, Bernard JK, West JW. Performance of lactating dairy cows fed ryegrass silage and corn silage with ground corn, steam-flaked corn, or hominy feed. J. Dairy Sci. 2009;92:1117–1123.

[56] Cooke KM, Bernard JK, West JW. Performance of dairy cows fed annual ryegrass silage and corn silage with steam-flaked or ground corn. J. Dairy Sci. 2008;91:2417–2422.

[57] P. Yu, J.T. Huber, F.A.P. Santos, J.M. Simas, C.B. Theurer. Effects of Ground, Steam-Flaked, and Steam-Rolled Corn Grains on Performance of Lactating Cows. Journal of Dairy Science, Volume 81, Issue 3, March 1998, Pages 777–783.

[58] Plascencia A, Zinn RA. Influence of flake density on the feeding value of steam-processed corn in diets for lactating cows. J. Anim. Sci. 1996;74:310–316.

[59] Zinn RA, Barajas R. Influence of flake density on the comparative feeding value of a barley-corn blend for feedlot cattle. J. Anim. Sci. 1997;75:904–909.

[60] Plascencia A, González-Vizcarra VM, López-Soto MA, May D, Pujol LC, Ruiz GF, Vega SH. Influence of cracked, coarse ground or fine ground corn on digestion, dry matter intake and milk yield in Holstein cows. J. Appl. Anim. Res. 2009;35:149–154.

Section 3

Herbivory Effects on Plant Communities

Chapter 6

Positive Indirect Interactions in Marine Herbivores and Algae

Jacqueline E. Bell and Jane E. Williamson

Additional information is available at the end of the chapter

http://dx.doi.org/10.5772/67343

Abstract

There is an increasing interest in how nested positive indirect interactions involving at least three species maintain community structure. Recent research shows that positive indirect effects can strongly influence community structure, organisation and functioning. It is thus important to understand and identify positive indirect effects for the purpose of predicting system responses to certain perturbations. In order to investigate indirect effects, experimental manipulations must be carried out within the entire framework of the community of interest. Hence, often due to logistical difficulties, indirect effects, especially those that yield positive results, have been less studied. Here we present a synthesis of current information on patterns of positive indirect effects and review and compare recently conducted experimental studies in marine herbivores and algae.

Keywords: indirect, plant, cascade, habitat, facilitation

1. Introduction

In this chapter, we synthesise current information and case examples of defined patterns of positive indirect interactions in marine herbivores. These types of interactions occur when one species causes a change in a second species, which successively affects a third species and where at least one species is benefited and neither is harmed [1]. Herbivores in marine ecosystems have the ability to drastically modify the biogenic structure of habitats. To date, most of the ecological literature on marine herbivory has focused on negative effects arising from the overharvest of predators or shifting environmental conditions, which can lead to a loss of structural habitat. This chapter highlights the diverse roles that herbivorous grazers can play in directly and indirectly enhancing species diversity. The importance of multispecies interactions involving herbivores has recently been recognised. We highlight that greater survivorship of contributing species inside such associations, as well as behavioural habitat selection, is important in the establishment of such interactions and that food provision is an important driver in their maintenance in marine systems. This chapter concludes with an emphasis on the importance of understanding multispecies interactions in successful management of marine ecosystems. In order to accurately predict the impact of potential perturbations and for mitigation of effects, future research should refocus on the entire ecosystem framework to capture potentially important positive indirect effects that might further define species relationships.

1.1. Positive interactions

The importance of positive interactions between species is increasingly acknowledged in contemporary ecological theory [1–3]. Such interactions occur between two or more species when at least one participant benefits and neither are harmed and take place, simply, as either commensal (+, 0) or mutual effects (+, +) [1, 2, 4, 5]. A species that has a positive effect on another is referred to as a facilitator [2, 6]. Facilitative or positive interactions tend to be most common in environments with high physical stress and/or where strong consumer pressure exists [3, 4]. Here, facilitators play a positive role by ameliorating environmental stress and by creating complex habitat that can lessen the effects of competition and/or predation [1]. Relationships between facilitators and associated organisms may be obligate or facultative, depending on the level of risk to survival for the associated species outside of the relationship [4].

1.2. Direct and indirect interactions

Interactions between species can be both direct and indirect, yielding both positive and negative results [7–9]. A direct effect occurs as a result of a physical interaction between two species [10] and includes processes such as predation, interference competition, inhibition of recruitment, inhibition of feeding, enhancement of recruitment and provision of habitat or shelter [8]. Indirect effects occur in multispecies assemblages when the action of one species causes a change in a second species, subsequently impacting on a third species [11, 12]. This type of interaction includes processes such as keystone predation, tritrophic interactions, exploitation competition, apparent competition, indirect mutualism, indirect commensalism, habitat facilitation and associational resistance [13].

Indirect effects occur when a species is involved in a series of strong pairwise interactions that are not independent of other species [13]. Indirect effects generally occur in a system via two ways [13]. The first is referred to as an interaction chain where species C indirectly changes the abundance of species A by changing the abundance of an intermediary species, species B, which interacts with both [13]. The second is termed either interaction modification or higher order modification and occurs more commonly. It occurs when the abundance of species C changes, causing an indirect effect on the abundance of species A by affecting the interaction between species A and species B [13] (**Figure 1**).

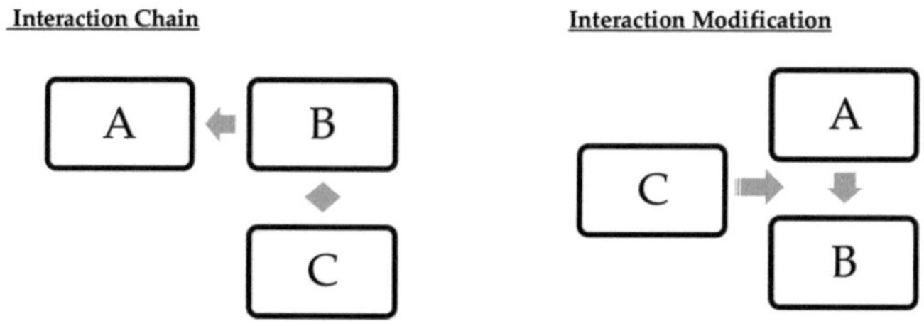

Figure 1. Two fundamental ways in which indirect effects can occur within an ecosystem (adapted from [13]).

Indirect effects also arise through changes in a physical and/or chemical component of the environment, as well as through another species [14]. For example, the effects of nutrient addition to a plant-endophage-parasitoid trophic chain can result in two types of indirect effects [15]. Fertilised plants (*Chromolaena squalid*) produce larger flower heads that act as a shelter against endophageous insects [15], representing an interaction modification [16]. Concurrently, this fertilisation results in an increase in the nutritional quality of the plant, which in turn increases the quality of the endophages as hosts to parasites [15] representing an interaction chain [16]. A similar example involves fish foraging, which causes a direct increase in sedimentation. This in turn has an effect on the abundance of invertebrates consuming primary producers [17–19]. Chemical cues and chemical communication can also indirectly mediate behaviour and strongly affect community structure [20]. For example, when given a choice between different cue sources, fish (*Lipophrys pholis*) and crabs (*Carcinus maenas*) that consume snails (*Littorina obtusata*) are more attracted to algae (*Ascophyllum nodosum*) that have recently been grazed to algae that have not [21]. It is thought that this indirect effect may have evolved in algae as a mechanism for protection against attacking herbivores [20].

Indirect effects within ecosystems may have important implications. It is thus important to understand and identify such effects for the purpose of predicting system responses to certain perturbations [13]. For example, human-induced perturbations to environments, such as replacement of natural marine habitats with artificial structures such as piling, marinas and seawalls, can have extensive direct and indirect repercussions on the abundances of biota within the ecosystem, and it is important for us to be able to identify such processes [22]. Environmental impacts such the introduction, increase, reduction or extinction of species can have widespread repercussions for the rest of an ecosystem [23]. Categorisation of organisms into separate trophic levels according to their feeding preference provides a useful foundation with which to understand ecological systems [23]. Relationships between producers and consumers can be examined in this way to determine which trophic level, if removed, may control community composition [24].

Detection of indirect effects is, however, sometimes more complex than this, as indirect effects can be masked as direct effects within manipulation experiments. For example, when avian predation pressure was experimentally manipulated within an intertidal community, both direct and indirect effects were found [25]. An increase in predatory gulls reduced the density of the limpet *Lottia digitalis* [10]. The seemingly direct effect of foraging gulls on limpet abundance was later found to be partial due to an indirect effect involving a change in the abundance of the cryptic goose barnacle (*Pollicipes polymerus*), which comprised the habitat in which the limpet (*L. digitalis*) preferentially colonised [10]. As a direct result of gull predation, the area covered by the cryptic goose barnacle was dramatically reduced, thus increasing the area covered by the habitat-forming mussel, *Mytilus californianus*. A reduction in the preferred cryptic habitat meant an increased risk of predation for *L. digitalis* and thus a reduction in its abundance [6]. This released the limpet *L. strigatella* from exploitative competition with *L. digitalis*, and thus an increase in the abundance of the former was observed [10]. Results of this experiment reveal that gull predation, in fact, indirectly decreases the abundance of the limpet *L. digitalis*, which in turn increases the abundance of the limpet *L. strigatella*, via a decrease in the preferred cryptic habitat of *L. digitalis*, causing a reduction in the strength of exploitative competition between the two species [10]. This example demonstrates the importance of long-term experimental manipulations that consider the full complexities of the community of interest, for the purpose of detecting the underlying indirect effects. It also shows that conclusions from short-term experimental manipulations that simplify systems to direct interactions between species pairs can give questionable results [25].

Many direct effects within marine communities have been investigated in detail. Indirect effects, however, especially those that yield positive results, are less studied [10, 11]. The majority of indirect effects have been inferred from manipulative experiments that were designed to test other interactions rather than having been tested directly (e.g., [8, 26]). This may be due to the logistic difficulties in observing indirect effects within the marine environment or the difficulty in distinguishing between the effects of indirect and direct processes within multispecies interactions [8, 11, 12]. Nevertheless, there is little doubt that positive indirect effects are more common than historically thought and a growing body of work has revealed the importance of such effects within marine communities. Whilst there are almost an infinite number of associations involving indirect interactions between organisms, this chapter focuses on the current trends and significance of positive indirect effects that have shown to be ecologically important within benthic marine communities.

1.3. Patterns of positive indirect interactions associated with marine herbivores within marine communities

1.3.1. Food webs and trophic cascades

Food webs are crucial elements of community ecology as they describe the flow of energy and materials from one trophic (consumer) level to another [7, 8, 24, 27–30]. Species interactions within food webs are important when considering species demography and community structure across different habitats [23, 24]. In several cases, removal or introduction of a predatory trophic level can cause a cascading effect on other trophic levels [7, 10, 24, 31–34]. Such trophic cascades are simple indirect effects that occur as a result of consumer-resource interactions [13]. The most studied and classic marine example is the north-eastern Pacific trophic cascade involving sea otters, sea urchins and kelp [32]. Revival of the sea otter *Enhydra lutris* population had positive indirect effects on the near-shore benthic community structure [32] via a decrease in sea urchin *Strongylocentrotus polyacanthus* herbivory, which in turn caused an increase in kelp *Laminaria* spp. cover and habitat, as well as changes to the physical parameters of the environment (e.g., water flow, light penetration) [32, 35, 36].

The potential for human-induced trophic cascades has become more apparent in recent years [9, 34]. Introduction of 'no take' marine reserves has reduced the impacts of humans on predatory levels in specific areas, resulting in positive indirect effect within these marine communities than can be observed for the first time [37]. A reversal in community structure was observed within Leigh Marine Reserve in New Zealand as a result of the elimination of fishing since 1976 [37]. Herbivory and the density of sea urchins declined with an increase in predation, which in turn increased the biomass of primary producers and altered seaweed community structure [37]. When comparisons were made between the area within the reserve and the area adjacent to the reserve for the 4–6 m depth zone, a marked distinction could be made between urchin-induced barrens (areas devoid of kelp) as the dominant habitat outside the reserve and the complex kelp habitat that was dominant within the reserve [37].

Predator diversity can strengthen positive trophic cascades by further reducing herbivory and increasing plant biomass [38]. Interspecific competition among predators is considered pivotal in maintaining food web dynamics, community structure and ecosystem functioning within marine systems [38–40]. For instance, an increase in predator diversity is believed to increase the likelihood of keystone predation or facilitation within the predatory assemblage, thus enhancing the efficiency of prey consumption [41]. Predators can affect plant biomass through 'density-mediated indirect interactions' (DMII), by reducing herbivore abundances, or through 'trait-mediated indirect interactions' (TMII) by reducing parameters such as the foraging period of herbivores [42]. Interestingly, Bruno and O'Connor [34] found that inclusion of omnivores in predator assemblages could reverse predicted positive indirect relationships between predator diversity and plant biomass. Through direct consumption of algae, omnivores effectively by-passed the trophic cascade. Thus, the magnitude and direction of changes in this community structure were due to changes in predator diversity. Cascades can sometimes be difficult to predict due to the multiple counteracting interactions that occur, especially when more generalist feeders like omnivores are included [38]. A review by Duffy et al. [31] came to a similar conclusion. Whilst horizontal predator diversity has indirect effects on primary production, the strength and sign of such effects depend on the diversity of prey types consumed (omnivore versus predator) and of course prey behaviour [43].

1.3.2. Indirect mutualisms

Indirect mutualisms can be defined as the shared indirect positive effects that one species has on another [44, 45]. They occur when the benefit exceeds the cost for both participants within an interspecific interaction (+, +) [46]. Positive interactions within the marine environment, especially mutualisms, are surprisingly widespread and play a critical role in shaping ecosystems [5]. Indirect mutualisms can arise through a number of mechanisms but typically involve a consumer-resource interaction linked with competitive interactions and are more likely to occur if the competitive relationship between resource species is strong [13]. In the presence of a competitive hierarchy between resource species, the interaction may become a direct commensalism (+, 0) [47].

Foundation species provide structure to the community and include groups such as kelp, coral and seagrass [5]. Mutualistic interactions frequently occur between foundation species and their residents whereby both resident and foundation species benefit [5]. This process, also known as indirect facilitation [1], will be discussed in more detail later in this chapter. Perhaps the most well-studied mutualistic interaction involving a foundation species within a marine community is that between corals and their photosynthetic dinoflagellate symbionts, zooxanthellae [5]. Photosynthesis by zooxanthellae provides the coral host with carbohydrates, whilst the resident zooxanthellae receive nutrients via nitrogenous waste from the prey of their carnivorous coral host [5]. The carbohydrates are used by the coral for calcification and growth, allowing them to grow at a rapid rate, which is necessary for survival [5]. Whether such rapid growth will be enough to ensure coral survival in many regions under rapid sea level, change is still unknown. Survival of one of the most biologically diverse ecosystems in the world would certainly be severely compromised without this mutualistic interaction [5].

Corals persist in tropical environments due, in part, to the efficient grazing activity of herbivores that prevent overgrowth by fouling algae [48]. Within temperate marine communities, however, fewer species of coral survive due to the competitive advantage that algae have over corals, where herbivory is less intense [48]. Contrary to this trend, the coral *Oculina arbuscula* persists in temperate waters off North Carolina despite the prevalence of macroalgae due to a mutualistic relationship with the omnivorous crab *Mithrax forceps* [46]. The coral harbours the crab, which consumes all types of algae and invertebrates inhabiting the coral. The crab uses the coral for protection from predators and gains a dietary advantage from the coral by consuming the lipid-rich coral mucus [48]. This mucus may also attract symbionts that further protect the coral from predation [48].

A negative consumer-resource interaction can flip to a positive interaction through changes to mutualistic effects [43]. Coralline algae, for example, are typically consumed by molluscs that scrape them from the rocks they inhabit with their hardened radulae [49]. Within the Belize Barrier Reef, approximately half of the diet of the herbivorous chiton, *Chonoplax lata*, is made up of its preferred coralline algal host *Porolithon pachydermum* [49]. Feeding by the chiton c burrows and excavates into the coralline algae, causing damage to the host [49]. When the chitons are experimentally removed, however, the coralline algae become extensively fouled by epiphytic algae, which attract deep biting by powerful herbivorous fish, including parrot fish. This form of herbivory causes substantially more damage to the coralline algae than that caused by the chiton [47]. Thus, removal of the chiton caused an increase in grazing damage rather than a decrease. Herbivorous damselfish can form similar mutualisms with algae. By protecting their food source, less grazing activity occurs to the algal mats on which they feed [50]. As a result, these algal mats are far more species rich and occur in greater biomass than those subjected to all types of grazing [50]. In fact, when damselfish are experimentally removed, these algal mats are consumed entirely within hours [50, 51].

Mutualists in one ecological context may be adversaries in another ecological context [5]. Whilst indirect mutualism yields positive results by definition, this type of effect is often linked with negative interactions, such as exploitative competition [13]. When two competing species are considered in a community context, the effects of a nearby competitor can sometimes counterbalance the negative effects of competition by lessening physical stresses or preventing attacks by enemies [5]. A classic example is where the addition of a seastar within an intertidal community directly decreases the abundance of the resident mussels (*Mytilus*), which in turn makes space for competitively inferior sessile species [52]. A similar example is described by Wulff [53] whereby particular species of sponges grew better when surrounded by other species of sponges than when grown with conspecifics or when grown alone. This is thought to be due to a nearby competitor lessening the impacts of predation, acting as a positive trade-off to the negative effects of competition [5].

Mutualistic interactions have long been considered a coevolved trait, involving species that are coupled consistently in space and time; however, this is not always the case [5]. Some interactions that appear to have coevolved do not have an obvious coevolutionary history [54, 55], suggesting that their occurrence may have arisen as an incidental benefit [56]. For example, damselfish seek refuge from predators by hiding within branching coral [5]. The damselfish benefits mutualistically the coral by providing nutrients whilst in hiding, via excretion, thus allowing the coral to grow at a faster rate [57]. Extensive branching on this type of coral is thought to have evolved in response to feeding and reproductive needs rather than to take up nutrients provided by the damselfish [5]. Similarly, growth of the brown encrusting alga *Pseudolithoderma* sp. is increased through uptake of ammonium by overlying live honeycomb barnacles (*Chamaesipho columna*) [58]. Occurrence of the alga on the barnacles is most likely due to a refuge from herbivory, and it is thought that the alga reduces the impact of desiccation for the barnacles during low tides [58, 59].

1.3.3. Associational resistance

Associational resistance occurs when an organism takes refuge from predation by associating with a habitat-forming competitor (+, +) or (+, 0) [60]. Palatable marine plants, for example, are more vulnerable to herbivory when occurring alone, but herbivory is reduced and growth enhanced, when the same species grows interspersed with algae that are unpalatable to herbivores [61–63]. This is a facilitative-commensalistic (+, 0) example of associational resistance whereby the palatable plant has a clear benefit by association; however, the unpalatable plant neither benefits nor suffers [1]. Such an interaction can become antagonistic (+, −) if the palatable plant outgrows the unpalatable plant, making the unpalatable plant more attractive to herbivory [1]. In this instance the relationship could also be considered parasitic [1]. When the unpalatable plant remains dominant in the community, however, species growth and diversity can increase significantly by providing a safe haven for the palatable species [63]. This example highlights the transient nature of some associations over time, such that interactions can flip from being positive to negative and potentially back again, given particular biotic and abiotic circumstances [63].

Mobile organisms, often herbivorous, can also take refuge from predation by association with seagrasses, kelps, corals and other sessile or less mobile organisms that provide structural and morphological defences [1]. Smaller marine invertebrates can shelter within the structurally complex habitat formed by seagrass, kelp and corals for protection from predators using their host as both food and habitat [1]. Whilst structural complexity can play a large role in providing safe havens from predation, the chemical makeup of plants can also deter larger consumers [1]. Some marine invertebrates inhabit plants that contain noxious antipredator chemicals and feed on species other than their host [1]. In such situations the benefit of refuge is thought to outweigh the importance of the quality of the food. For example, the juvenile sea urchin *Holopneustes purpurascens* inhabits the chemically defended foliose red alga *Delisea pulchra* [64]. *H. purpurascens* exhibits a diel pattern of movement on its host plant. It remains wrapped within its host during the day, when predation is greatest, and is more exposed at night, for purposes thought to include nutritional gain, reproduction, avoidance of photo damage and microenvironmental variation associated with the host alga [65]. When *H. purpurascens* reaches a certain size, it moves to a new host plant, the kelp *Ecklonia radiata*, on which it feeds [64]. At this point in its life history, it is thought that the benefit of a more nutritious and easily accessible food source outweighs the benefit of refuge via a chemically noxious host [64].

The decorative behaviour of certain crab species with chemically defended plants is a similar scenario. The decorator crab *Libinia dubia* camouflages itself by covering its carapace with the chemically noxious brown alga *Dictyota menstrualis* [66]. The diterpene alcohol produced by the brown alga deters predators by making the alga unpalatable [66]. The diterpene alcohol also acts as a cue for the crab to commence decorative behaviour [66]. Studies have shown that without this behavioural adaptation, *L. dubia* would most likely become extinct [66]. It is thought that the relationship between the decorator crab *L. dubia* and the brown alga *D. menstrualis* may well be mutualistic, whereby the alga benefits though and through reduced herbivory via the consumption of amphipods by the crab [66] and gains nutritionally via crab excretion, as in the relationship between the brown alga *Pseudolithoderma* sp. and the barnacle *Chamaesipho columna* [58].

Associational resistance can also occur between invertebrates. For example, less mobile sea urchins (*Parechinus angulosus*) provide a stable habitat for juvenile abalone that are at risk of predation by crayfish [67]. Experimental removal of urchins indirectly affected recruiting abalone by causing an increase in sediment. McClintock and Janssen [68] document a similar occurrence whereby an amphipod increased its chances of survival by capturing a chemically defended pteropod, effectively exploiting the pteropod's chemical defence for its own protection.

Associational resistance is sometimes considered facilitative when the species that provides the associational resistance is facilitated by the association. For example, an Antarctic sea urchin facilitates dispersal of chemically defended seaweeds that have become detached during storms [69]. The sea urchin exhibits a similar decorative behaviour where it collects reproductively viable individuals for camouflage to deter predation whilst also preventing the seaweed from being carried ashore or below the photic zone [69]. This example could also be defined as mutualistic.

1.3.4. Facilitation cascades

Facilitation cascade is another example of a positive indirect effect and is commonly observed in marine herbivores and macroalgae. Within a facilitation cascade, the basal habitat former facilitates an intermediate habitat former, which in turn facilitates a focal species. In marine environments, where predation is often intense and waves and currents produce abiotically stressful conditions, positive interactions among species, such as facilitation cascades, are expected to play a particularly important role in the structure and organisation of ecological communities [1, 4, 6, 70, 71].

Marine benthic communities inhabit highly dynamic environments [72]. Storm surges, wave action, tides and currents, as well as biotic factors related to food web dynamics; all contribute to the dynamics of this environment [73]. Facilitator species within these systems include benthic species such as kelps [24], seagrasses [74] and mangroves [75]. These mitigate environmental stressors for associated species through substrate formation [76, 77]; enhancement of larval settlement [78]; provision of food [79]; shelter from physical forces such as wave action, tides and currents [80]; and refuge from predation [81]. These species often form large aggregations whereby facilitation of generally smaller species, often herbivores, occurs through the creation of habitat heterogeneity [76].

Herbivores in marine ecosystems have the ability to drastically modify the biogenic structure of habitats. Sea urchins, for example, are major grazers in rocky reef ecosystems, often maintaining areas devoid of macroalgae, namely, 'urchin barrens' [82]. To date, most of the ecological literature has focused on the cascading negative effects of increasing herbivore abundance arising from the overharvest of their predators or shifting environmental conditions, which can lead to a loss of structural habitat [32, 83–87]. However, some herbivores can have positive effects on particular associated species. These positive effects most likely occur at smaller scales than the negative effects associated with large-scale herbivory and often within facilitation cascades, whereby complex systems of direct and indirect pathways make them more difficult to uncover.

Perhaps the most common and simplest way that a herbivore can mediate a facilitation cascade is by providing shelter for other small invertebrates [88–91]. In mangrove forests, for example, marine invertebrates such as sponges and barnacles are directly facilitated by the mangroves in which they inhabit and, in turn, indirectly facilitate the mangroves by providing physical barriers, thus protecting them from wood-boring isopods [92]. Within the lagoons of French Polynesia, gammarid amphipods and chaetopterid polychaetes induce the growth of branch-like 'fingers' on corals through nutrient provisioning, which in turn facilitate the abundance and diversity of fishes [93]. In intertidal cobblestone beaches, cordgrass beds provide habitat for mussels, which in turn create crevice space a shelter to an array of other marine invertebrates [77]. Thomsen [94] conceptualises a specific type of facilitation cascade, described as a 'habitat cascade'. This type of interaction is characterised when a basal habitat former, typically a large primary producer, creates space for an intermediate habitat former to live, that in turn creates habitat for the focal organism.

One example of a habitat cascade mediated by a marine herbivore is that between the common kelp *Ecklonia radiata*, the sea urchin *Holopneustes purpurascens* and the gastropod *Phasianotrochus eximius*. Within this relationship, the intermediary species, the short-spined urchin, *H. purpurascens,* uses its tube feet to wrap itself in the laminae of the kelp [36, 64]. It also preferentially consumes the kelp [95]. The focal organism, the gastropod *P. eximius,* resides with *H. purpurascens* in the temporary shelter the urchin builds within the fronds of the kelp [65]. The relationship is considered facultative, as *P. eximius* can survive in different types of habitats but is most abundant on *E. radiata* plants with *H. purpurascens* throughout the year [96]. Due to its small size, *P. eximius* is likely to be vulnerable to predation outside of its preferred complex habitat structure. The modified habitat in which both species exist is thought to benefit the sea urchin by providing it with a shelter from predation but also from abrasion by kelps and other objects 'whipping' by in the water due to adverse abiotic factors such as wave action, tides and currents [65].

Covering behaviour in other species of sea urchins has also been considered an adaptation to avoid surge [97]. The sea urchin *Toxopneustes roseus* covers itself in shell fragments and foliose algae in areas of high surge throughout the Gulf of California [97]. It is possible that *H. purpurascens* has adapted in a similar way to *T. roseus* by covering itself to mitigate wave action within the exposed environment in which it inhabits [36]. It is highly likely, therefore, that *P. eximius* also benefits from inhabiting the shelter built by *H. purpurascens*.

Impacts on one species within a facilitation cascade can profoundly change the balance of the relationship. Recently, *H. purpurascens* in this region has been associated with the outbreak of a disease caused by the opportunistic pathogen *Vibrio anguillarum* [98]. The disease reduces the capacity of the urchin to wrap algae around itself and ultimately leads to death of the urchin. The disease is water-borne, and prevalence of the disease is exacerbated by increases in water temperature, such as those associated with climate change [98]. Whilst the impact of the urchin disease on the health and demography of both kelp and gastropod is currently unknown, it is highly likely that both may suffer through prolonged contact with diseased urchins. *P. eximius* may also face reduced availability of habitat formed by *H. purpurascens* should the abundance of urchins be dramatically impacted.

Plants often mediate facilitation cascades. These interactions typically occur in temporally separated, spatially separated or taxonomically distinct species [99–101]. Thomsen [94] investigated one particular example whereby small herbivorous marine invertebrates facilitate habitat for seaweeds, which in turn facilitate habitat for focal species of invertebrates and epiphytes. Other examples involve two levels of plant facilitation. For example, the seaweed *Hormosira banksii* provides habitat for the obligate epiphyte *Notheia anomala,* which in turn facilitates species richness and diversity of mobile invertebrates [102]. Similarly, temperate Australian mangrove forests facilitate free-living algae, which in turn facilitate a dense and diverse assemblage of epifaunal molluscs [103].

For small marine herbivores, associations with larger, habitat-forming herbivores can be driven by a range of environmental obstacles that need to be efficiently overcome to survive [104, 105]. These not only include the need for shelter but also finding a reliable and nutritious food source and access to mates, the former two being generally considered the most important driving factors in habitat and/or host choice [79, 104–106]. Ideally, an individual will choose a habitat or host that provides all of these attributes [16].

By investigating both the direct and indirect effects of species interactions, often a seemingly simple association will be based on more complex foundations. For example, grazing sea urchins and gastropods are directly facilitated by mussel beds by feeding on attached algae;

the mussels are indirectly facilitated by the grazers that keep them free from algal growth and reduce the potential for mussel dislodgement by up to 30-fold [107]. Similarly, juvenile abalone that recruit to the underside of the sea urchin *Parechinus angulosus* [67] receive protection by the urchin but also provision of food via drift algae that the urchin captures on its spines for its own consumption [67]. Another example can be observed between the isopod *Dulichia rhabdoplastis* and sea urchin *Strongylocentrotus franciscanus*, which appears to be indirectly mediated [108]. Within this relationship, the isopod builds strings of detritus made from its own faecal pellets that it connects to the spines of the sea urchin [108]. The strings are colonised by a rich layer of diatoms, which the isopod subsequently consumes [108]. Here, the sea urchin indirectly facilitates the isopod by providing it with a habitat that it uses to capture its prey [108]. This species may also benefit directly by using the spines of the sea urchin as refuge when needed.

Facilitation cascades are not exclusive to herbivores. An invasion by non-native bullfrogs has been facilitated by the coevolved non-native sunfish, where the sunfish increased bullfrog tadpole survival by consuming dragonfly nymphs that preyed on the tadpoles [109]. Such an interaction between two non-native species also has the potential to exacerbate impacts of species invasion [109].

2. Conclusion

Positive interactions involving marine herbivores and algae have been increasingly recognised for their importance in the structure and functioning of ecosystems [94]. However, studies focusing on the role of negative species interactions in shaping ecosystems such as over harvest of predators or shifting environmental conditions, which can lead to loss of structural habitat, still far outweigh those focusing on the importance of positive effects [32, 83–87]. Herbivores in marine ecosystems have the ability to drastically modify the biogenic structure of habitats. Indirect effects add to the complexity with which ecosystems function and are intrinsically difficult to quantify, often requiring long-term and manipulative experiments [101]. Whilst interest in indirect effects has recently grown, there is still a gap in our understanding of the roles that individual indirect effects have and their importance within many systems [16]. An understanding of positive interactions, and both the direct and indirect pathways of occurrence, is essential to predict accurately the impact of potential perturbations for successful management of ecosystems. Greater survivorship of contributing species inside such associations as well as behavioural habitat selection is important in the establishment of such interactions, and food provision is an important driver in their maintenance in marine systems. Whilst difficult, future research should focus on the entire framework of these ecosystems to capture potentially important cascading effects that might further define species relationships. Experiments should centre on the effects of feeding behaviour and the nutritional benefits of association, the role of predation and the risks herbivores face beyond the association as well as environmental stressors such as wave action and climate change on the survival of associates within and outside of preferred habitats.

Throughout the past 50–100 years, human impacts on marine ecosystems (such as overfishing) have resulted in a downturn in the abundance of species that prey on herbivores in some areas [110]. Within such areas this has caused an increase in the abundance of herbivorous species and in turn is likely to have had a positive effect on species that associate with sea urchins [111]. Recently, however, direct threats on herbivores by humans, such as harvesting for food [112], creating suboptimal conditions that, increased sedimentation [113] and ocean acidification [114] on local to regional scales, have increased, which in turn will negatively impact on the species with which the herbivores facilitate. This issue has been identified as particularly relevant to commercially harvested species that rely on herbivore for survival, such as the abalone *H. midae*, which depends on the sea urchin *P. angulosus* throughout its juvenile stage for both food and shelter in South Africa. Depletion of sea urchin stocks in this location has seen a decline in abalone recruits, which have had significant impacts on the abalone industry in this region [67]. This chapter highlights the diverse roles that herbivorous grazers play in directly and indirectly enhancing species diversity. Unfortunately, however, the relatively unstudied nature of many species interactions within the marine environment means that many of these types of associations may disappear before we have the opportunity to understand their importance within ecosystem functioning. With a greater level of understanding of the important roles that herbivores play within various marine ecosystems, the cascading effects as a result of threats to herbivores can be managed appropriately, for the purpose of maintaining future biodiversity.

Author details

Jacqueline E. Bell[1,2] and Jane E. Williamson[2*]

*Address all correspondence to: jane.williamson@mq.edu.au

1 Boffa Miskell Ltd., Hamilton, New Zealand

2 Department of Biological Sciences, Macquarie University, Sydney, Australia

References

[1] Stachowicz, J. J. Mutualism, facilitation, and the structure of ecological communities. Bioscience, 2001; **51**:235–246.

[2] Bruno, J. F., Stachowicz, J. J. and Bertness, M. D. Inclusion of facilitation into ecological theory. Trends Ecol. Evol. 2003; **18**: 119–125.

[3] Silliman, B. R., Bertness, M. D., Altieri, A. H., Griffin, J. N., Cielo Bazterrica, M., Hidalgo, F.J., Crain, C.M., Reyna, M.V. Whole-community facilitation regulates biodiversity on Patagonian rocky shores. PLoS One. 2011; **6**: e24502.

[4] Bertness, M. D., Callaway, R. Positive interactions in communities. Trends Ecol. Evol. 1994; **9**: 191–193.

[5] Hay, M. E., Parker, J. D., Burkepile, D. E., Caudill, C. C., Wilson, A. E., Hallinan, Z. P., and Chequer, A. D. Mutualisms and aquatic community structure: the enemy of my enemy is my friend. Ann. Rev. Ecol. Evol. Syst. 2004; **35**: 175–197.

[6] Hacker, S. D., and Gaines, S. D. Some implications of direct positive interactions for community species diversity. Ecology. 1997; **78**: 1990–2003.

[7] Carpenter, S. R., Kitchell, J. F., Hodgson, J. R. Cascading trophic interactions and lake productivity. Bioscience. 1985; **35**: 634–639.

[8] Menge, B. A. Indirect effects in marine intertidal interaction webs: patterns and importance. Ecol. Monograph. 1995; **65**: 21–74.

[9] Polis, G. A., Scars, A. L. W., Huxel, G. R., Strong, D. R., and Maron, J. When is a trophic cascade a trophic cascade? Trends Ecol. Evol. 2000; **15**: 473–475.

[10] Wootton, T. J., Indirect effects, prey susceptibility, and habitat selection: impacts of birds on limpets and algae. Ecology. 1992; **73**: 981–991.

[11] Dill, L. M., Heithaus, M. R., Walters, C. J. Behaviourally mediated indirect interactions in marine communities and their conservational implications. Ecology. 2003; **84**: 1151–1157.

[12] Schmitt, R. J. Indirect interactions between prey: apparent competition, predator aggregation, and habitat segregation. Ecology. 1987; **68**: 1887–1897.

[13] Wootton, T. J. The nature and consequences of indirect effects in ecological communities. Annu. Rev. Ecol. Syst. 1994; **25**: 443–466.

[14] Strauss, S. Y. Indirect effects in community ecology: their definition, study and importance. Trends Ecol. Evol. 1991; **6**: 206–210.

[15] Dias, A. T., Trigio, J. R., Lewinsohn T. M. Bottom-up effects on a plant-endophage-parasitoid system: the role of flower-head size and chemistry. Aust. Ecol. 2010; **35**: 104–115.

[16] Wootton, T. J. Indirect effects in complex ecosystems: recent progress and future challenges. J. Sea Res. 2002; **48**: 157–172.

[17] Flecker, A. S. Fish trophic guilds and the structure of a tropical stream: weak direct vs strong indirect effects. Ecology. 1992; **73**: 927–940.

[18] Power, M. E. Resource enhancement by indirect effects of grazers: armoured catfish, algae and sediment. Ecology. 1990; **71**: 897–904.

[19] Pringle, C. M., Blake, G. A., Covich, A. P., Buxby, K. M., Finlay, A. Effects of omnivorous shrimp in a montane tropical stream: sediment removal, disturbance of sessile invertebrates and enhancement of understory algal biomass. Oecologia. 1993; **93**: 1–11.

[20] Hay, M. E. Marine chemical ecology: chemical signals and cues structure marine populations, communities, and ecosystems. Annu. Rev. Mar. Sci. 2009; **1**: 193–212.

[21] Coleman, R. A., Ramchunder, S. J., Davies, K. M., Moody, A. J., Foggo, A. Herbivore induced infochemicals influence foraging behaviour in two intertidal predators. Oecologia. 2007; **151**: 454–463.

[22] Marzinelli, E. M., Zagal, C. J., Chapman, M. G., Underwood, A. J. Do modified habitats have direct or indirect effects on epifauna? Ecology. 2009; **90**: 10, 2948–2955.

[23] Connell, S. D., Vanderklift, M. A. Negative interactions: the influence of predators and herbivores on prey and ecological systems. in Connell, S. D., Gillanders, B. M. (Eds.) Marine Ecology. Melbourne: Oxford University Press; 2007.

[24] Davenport, A. C., Anderson, T. W. Positive indirect effects of reef fishes on kelp performance: the importance of mesograzers. Ecology. 2007; **88**: 1548–1561.

[25] Wootton, T. J. Direct and indirect effects of nutrients on intertidal community structure: variable consequences of seabird guano. J. Exp. Mar. Biol. Ecol. 1991; **151**: 139–153.

[26] Pennings, S. C. 'Indirect interactions on coral reefs' in Birkeland, C. (Ed) Life and Death of Coral Reefs. New York, USA: Chapman and Hall; 1994.

[27] Paine, R. T. Food webs: linkage, interaction strength and community infrastructure. J. Anim. Ecol. 1998; **49**: 667–685.

[28] Bascompte, J., Melian, C. J., Sala, E. Interaction strength combinations and overfishing of a marine food web. Proc. Natl. Acad. Sci. U. S. A. 2005; **102**: 5443–5447.

[29] Forrester, G. E., Dudley, T. L., Grimm, N. B. Trophic interactions in open systems: effects of predators and nutrients on stream food chains. Limnol. Oceanogr. 1999; **44**: 1187–1197.

[30] Nielsen, K. J. Bottom up and top down forces in tide pools: test of a food chain model in an intertidal community. Ecol. Monograph. 2001; **71**: 187–217.

[31] Duffy, E. J., Cardinale, B. J., France, K. E., McIntyre, P. B., Thebault, E., and Loreau, M. The functional role of biodiversity in ecosystems: incorporating trophic complexity. Ecol. Lett. 2007; **10**: 522–538.

[32] Estes, J. A., Smith, N. S., Palmisano, J. F. Sea otter predation and community organisation in the western Aleutian Islands, Alaska, Ecology. 1978; **59**: 822–833.

[33] Polis, G. A. Complex trophic interactions in deserts: an empirical critique of food web theory. Am. Nat. 1991; **138**: 123–155.

[34] Pace, M. L., Cole, J. J., Carpenter, S. R., Kitchel, J. F. Trophic cascades revealed in diverse ecosystems. Trends Ecol. Evol. 1999; **14**: 235–240.

[35] Estes, J. A., Palmisano, J. F. Sea otters: their role in structuring benthic nearshore communities. Science. 1974; **185**: 1058–1060.

[36] Steinberg, P. D. Interactions between the canopy dwelling echinoid *Holopneustes purpurascens* and its host kelp *Ecklonia radiata*. Mar. Ecol. Prog. Ser. 1995; **127**: 169–181.

[37] Shears, T., Babcock, R. C. Continuing trophic cascade effects after 25 years of no-take marine reserve protection. Mar. Ecol. Prog. Ser. 2003; **246**: 1–16.

[38] Bruno, J. F., O'Connor, M. Cascading effects of predator diversity and omnivory in a marine food web. Ecol. Lett. 2005; **8**: 1048–1056.

[39] Duffy, J. E. Biodiversity and ecosystem function: the consumer connection. Oikos. 2002; **99**: 201–219.

[40] Sih, A., Englund, G., Wooster, D. Emergent impacts of multiple predators on prey. Trends Ecol. Evol. 1998; **13**: 350–355.

[41] Ives, A. R., Cardinale, B. J., Snyder, W. E. A synthesis of subdisciplines: predator-prey interactions, and biodiversity and ecosystem functioning. Ecol. Lett. 2005; **8**: 102–116.

[42] Trussell, G. C., Ewanchuk, P. J., Bertness, M. D. Field evidence of trait mediated indirect interactions in a rocky intertidal food web. Ecol. Lett. 2002; **5**: 241–245.

[43] Duffy, J. E., Richardson, J. P., France, K. E. Ecosystem consequences of diversity dependant on food chain length in estuarine vegetation. Ecol. Lett. 2005; **8**: 301–309.

[44] Levine, S. H. Competitive interactions in ecosystems. Am. Nat. 1976; **110**: 903–910.

[45] Vandermeer, J. Indirect mutualism: variations on a theme by Stephen Levine. Am. Nat. 1980; **116**: 441–448.

[46] Bronstein, J. L. Conditional outcomes in mutualistic interactions. Trends Ecol. Evol. 1994; **9**: 214–217.

[47] Dethier, M. N., Duggins, D. O. An "indirect commensalism" between marine herbivores and the importance of competitive hierarchies. Am. Nat. 1984; **124**: 205–219.

[48] Stachowicz, J. J., Hay, M. E. Geographic variation in camouflage specialization by a decorator crab. Am. Nat. 1999; **156**: 59-71.

[49] Littler, M. M., Littler, D. S., Taylor, P. R. Selective herbivory increases biomass of its prey: chiton-coralline reef-building associations. Ecology. 1995; **76**: 1666–1681.

[50] Hixon, M. A., Bronstroff, W. N. Succession and herbivory: effects of differential fish grazing on Hawaiian coral-reef algae. Ecol. Monograph. 1996; **66**: 67–90.

[51] Ceccarelli, D. M., Jones, G. P., and McCook, L. J. Territorial damselfish as determinants of the structure of benthic communities on benthic reefs: an annual review. Oceanogr. Mar. Biol. 2001; **39**: 355–389.

[52] Paine, R. T. Intertidal community structure: experimental studies on the relationship between a dominant competitor and its principal predator. Oecologia. 1974; **15**: 93–120.

[53] Wulff, J. L. Mutualisms among species of coral reef sponges. Ecology. 1997; **78**: 146–159.

[54] Steneck, R. S. 'Plant-herbivore coevolution: a reappraisal from the marine realm and its fossil record' in John, D. M., Hawkins, S. J., and Price, J. H. (Eds) Plant-Animal Interactions in the Marine Benthos. Clarendon: Oxford; 1992.

[55] Vermeij, G. J. Time of origin and biogeographical history of specialised relationships between northern marine plants and herbivorous molluscs. Evolution. 1992; **46**: 657–664.

[56] Connor, R. C. The benefits of mutualism – a conceptual framework. Biol. Rev. 1995; **70**: 427–457.

[57] Liberman, T., Genin, A., Loya, Y. Effects of growth and reproduction of the coral *Stylophora pistillata* by the mutualistic damselfish *Dascyllus marginatus*. Mar. Biol. 1995; **121**: 741–746.

[58] Williamson & Rees Nutritional interaction in an Alga-Barnacle Association. Oecologia. 1994; **99**:16–20.

[59] Williamson, J. E., Creese, R. G. Colonisation and persistence of patches of the crustose brown alga *Pseudolithoderma sp*. J. Exp. Mar. Biol. Ecol. 1996; **203**: 191–208.

[60] White, J. A., and Andow, D. A. Habitat modification contributes to associational resistance between herbivores. Oecologia. 2006; **148**: 482–490.

[61] Poore, A. G. B., Hill, N. A. Spatial associations among palatable and unpalatable macroalgae: a test of associational resistance with a herbivorous amphipod. J. Exp. Mar. Biol. Ecol. 2005; **2**: 207–216.

[62] Poore, A. G. B. Spatial associations among algae affect host use in a herbivorous marine amphipod. Oecologia. 2004; **1**: 104–112.

[63] Hay, M. E. Associational plant defences and the maintenance of species diversity: turning competitors into accomplices. Am. Nat. 1986; **128**: 617–641.

[64] Williamson, J. E., Carson, D. G., de Nys, R., Steinberg, P. D. Demographic consequences of an ontogenetic shift by a sea urchin in response to host plant chemistry. Ecology. 2004; 85: 1355–1371.

[65] Rogers, C. N., Williamson, J. E., Carson, D. G., Steinberg, P. D. Diel vertical movement by mesograzers on seaweeds. Mar. Ecol. Prog. Ser. 1998; **166**: 301–305.

[66] Stachowicz, J. J., Hay, M. E. Reducing predation through chemically mediated camouflage: indirect effects of plant defences on herbivores. Ecology. 1999; **80**: 495–509.

[67] Day, E., Branch, G. H. Effects of sea urchins (*Parechinus angulosus*) on recruits and juveniles of abalone (*Haliotis midae*). Ecol. Monograph. 2002; **71**: 143–199.

[68] McClintock, J. B., Janssen, J. Pteropod abduction as a chemical defence in a pelagic Antarctic amphipod. Nature. 1990; **346**: 462–464.

[69] Amsler, C. D., McClintock, J. B., Baker, B. J. An Antarctic feeding triangle: defensive interactions between macroalgae, sea urchins and sea anemones. Mar. Ecol. Prog. Ser. 1999; **183**: 105–114.

[70] Jones, C. G., Lawton, J. H., Schachak, M. Organisms as ecosystem engineers. Oikos. 1994; **69**: 373–386.

[71] Jones, C. G., Lawton, J. H., Sachak, M. Positive and negative effects of organisms as physical ecosystem engineers. Ecology. 1997; **7**:1946–1957.

[72] Sousa, W.P. Natural disturbance and the dynamics of marine benthic communities. In: Bertness, M.D., Gaines, S.D. , Hay, M.E. (Ed) Marine ecological communities. Sinauer, Sunderland, 2001, pp. 85–130.

[73] Schiel, D. R. Rivets or bolts? When single species count in the function of temperate rocky reef communities. J. Exp. Mar. Biol. Ecol. 2006; 338: 233–252.

[74] Pagès, J. F., Gera, A., Romero, J., Farina, S., Garcia-Rubies, A., Hereu, B., Alcoverro, T. The Mediterranean benthic herbivores show diverse responses to extreme storm disturbances. PLoS One. 2013; **8**: e62719.

[75] Sheridan, P., Hays, C. Are mangroves nursery habitat for transient fishes and decapods? Wetlands. 2003; **2**: 449–458.

[76] Bruno J. F., Bertness M. D. Habitat modification and facilitation in Benthic Marine Communities. In Bertness M. D., Gaines S. D. and Hay M. E. (Eds.) Marine Community Ecology. Sinauer Associates: Sunderland, 2000, pp. 201–228.

[77] Altieri, A.H., Silliman, B.R., Bertness, M.D. Hierarchical organization via a facilitation cascade in intertidal cordgrass bed communities. Am. Nat. 2007; **2**: 195–206

[78] Rule, M. J., Smith, S. D. Spatial variation in the recruitment of benthic assemblages to artificial substrata. Mar. Ecol. Prog. Ser. 2005; **290**: 67–78.

[79] Boström, C., Mattila, J. The relative importance of food and shelter for seagrass-associated invertebrates: a latitudinal comparison of habitat choice by isopod grazers. Oecologia. 1999; **120**: 162–170.

[80] Kemppainen, P., van Nes, S., Ceder, C., Johannesson, K. Refuge function of marine algae complicates selection in an intertidal snail. Oecologia. 2005; **143**: 402–411.

[81] Grabowski, J. H., Hughes, A. R., Kimbro, D. L. Habitat complexity influences cascading effects of multiple predators. Ecology. 2008; **89**: 3413–3422.

[82] Steneck, R. S. Sea Urchins as drivers of shallow benthic marine community structure. Sea Urchins: Biol. Ecol. 2013; **38**: 195.

[83] Harrold, C., Reed, D. C. Food availability, sea urchin grazing, and kelp forest community structure. Ecology. 1985; **66**: 1160–1169.

[84] Carter, S. K., VanBlaricom, G. R., Allen, B. L. Testing the generality of the trophic cascade paradigm for sea otters: a case study with kelp forests in northern Washington, USA, Hydrobiologia. 2007; **579**: 233–249.

[85] Pinnegar, J. K., Polunin, N. V. C., Francour, P., Badalamenti, F., Chemello, R., Harmelin-Vivien, M. L., Hereu, B., Milazzo, M., Zabala, M., D'Anna, J., Pipitone, C. Trophic cascades in benthic marine ecosystems: lessons for fisheries and protected-area management. Environ. Cons. 2000; **27**: 179–200.

[86] Terborgh, J., Estes, J. A. (Eds.) Trophic cascades: predators, prey, and the changing dynamics of nature. Island Press, Washington, DC, USA, 2010.

[87] Byrnes, J., Stachowicz, J. J., Hultgren, K. M., Randall Hughes, A., Olyarnik, S. V., Thornber, C. S. Predator diversity strengthens trophic cascades in kelp forests by modifying herbivore behaviour. Ecol. Lett. 2006; **9**: 61–71.

[88] Dix, T. G. Association between the echinoid *Evechinus chloroticus* (Val.) and the clingfish *Dellichthys morelandi* Briggs. Pac. Sci. 1969. **23**; 332–336.

[89] Patton, W. K., Patton, R. J., Barnes, A. On the biology of *Gnathophylloides mineri* a shrimp inhabiting the sea urchin *Tripneustes ventricosus*. J. Crust. Biol. 1985; **5**: 616–626.

[90] Roberts, C. M., Hawkins, J. P. Extinction risk in the sea. Trends. Ecol. Ecol. 1999; **14**: 241–246.

[91] Kolm, N., Berglund, A. Wild populations of a reef fish suffer from the "nondestructive" aquarium trade fishery. Conserv. Biol. 2003; **17**: 910–914.

[92] Ellison A. M., Farnesworth, F. J. The ecology of Belizean mangrove-root fouling communities. I. Epibenthic fauna are barriers to isopod attack of mangrove roots. J. Exp. Mar. Biol. Ecol. 1990; **142**: 91–104.

[93] Bergsma, G. S. Coral mutualists enhance fish abundance and diversity through a morphology-mediated facilitation cascade. Mar. Ecol. Prog. Ser. 2012; **451**: 151–161.

[94] Thomsen, M. S. Experimental evidence for positive effects of invasive seaweed on native invertebrates via habitat-formation in a seagrass bed. Aquat. Invasion. 2010; **5**: 341–346.

[95] Williamson J. E., Steinberg P. D. Fitness benefits of size-dependent diet switching in a marine herbivore. Mar. Biol. 2012; **159**: 1001–1010

[96] Bell J. E., Bishop M. J., Taylor R. B., Williamson J. E. Facilitation cascade maintains a kelp community. Mar. Ecol. Prog. Ser. 2014; **501**: 1–10

[97] James, D. W. Diet, movement, and covering behavior of the sea urchin *Toxopneustes roseus* in rhodolith beds in the Gulf of California, México, Mar. Biol. 2000; **137**: 913–923.

[98] Sweet, M., Bulling, M., Williamson, J. E. New disease outbreak affects two dominant sea urchin species associated with Australian temperate reefs. Mar. Ecol. Prog. Ser. 2016; **551**: 171–183.

[99] Damman H. Patterns of interaction among herbivore species. In: Stamp N. E. and Casey T. M. (Eds.) Caterpillars: ecological and evolutionary constraints on foraging. Chapman and Hall, New York, 1993, pp. 132–169.

[100] Martinsen, G. D., Floate, K. D., Waltz, A. M., Wimp, G. M., and Whitham, T. G. Positive interactions between leafrollers and other arthropods enhance biodiversity on hybrid cottonwoods. Oecologia, 2000; **123**, 82–89.

[101] Ohgushi, T. Indirect interaction webs: herbivore-induced effects through trait change in plants. Ann. Rev. Ecol. Evol. Syst. 2005; **36**: 81–105.

[102] Thomsen, M. S., Metcalfe, I., South, P., Schiel, D. R. A host-specific habitat former controls biodiversity across ecological transitions in a rocky intertidal facilitation cascade. Mar. Freshw. Res. 2016; **67**: 144–152.

[103] Bishop, M. J., Fraser, J., Gribben, P. E. Morphological traits and density of foundation species modulate a facilitation cascade in Australian mangroves. Ecology. 2013; **94**: 1927–1936.

[104] Poore, A. G., Steinberg, P. D. Preference-performance relationships and effects of host plant choice in an herbivorous marine amphipod. Ecol. Monograph. 1999; **69**: 443–464.

[105] Orav-Kotta, H., Kotta, J. Food and habitat choice of the isopod *Idotea baltica* in the north-eastern Baltic Sea. Hydrobiologia. 2004; **514**: 79–85.

[106] Alfaro, A. C., Dewas, S. E., Thomas, F. Food and habitat partitioning in grazing snails (*Turbo smaragdus*), northern New Zealand. Estuar. Coast. 2007; **30**: 431–440.

[107] Witman, J. D. Subtidal coexistence: storms, grazing, mutualism, and the zonation of kelp and mussels. Ecol. Monograph. 1987; **57**: 167–187.

[108] McCloskey, L. R. A new species of Dulichia (*Amphipoda, Podoceridae*) commensal with a sea urchin. Pac. Sci. 1970; **24**: 90–98.

[109] Adams, M. J., Pearl, C. J., Bury, R. B. Indirect facilitation of an anuran invasion by non-native fish. Ecol. Lett. 2003; **6**: 343–351.

[110] Tegner, M. J., Dayton, P. K. Ecosystem effects of fishing in kelp forest communities. J. Mar. Sci. 2000; **57**: 579–589.

[111] Steneck, R. S., Graham, M. H., Bourque, B. J., Corbett, D., Erlandson, J. M., Estes, J. A., Tegner, M. J. Kelp forest ecosystems: biodiversity, stability, resilience and future. Environ. Cons. 2002; **29**: 436–459.

[112] Lawrence, J. M. (Ed.). Sea Urchins: Biology and Ecology. Elsevier Science, Tampa, Florida, USA, 2006.

[113] Phillips, N. E., Shima, J. S. Differential effects of suspended sediments on larval survival and settlement of New Zealand urchins *Evechinus chloroticus* and abalone *Haliotis iris*. Mar. Ecol. Prog. Ser. 2006; **314**: 149–158.

[114] Schlegel, P., Havenhand, J. N., Gillings, M. R., Williamson, J. E. Individual variability in reproductive success determines winners and losers under ocean acidification: a case study with Sea Urchins. PLoS one. 2012; **7**: e53118.

Chapter 7

Herbivory by Lizards

Carolina Pietczak and Lucas R. Vieira

Additional information is available at the end of the chapter

http://dx.doi.org/10.5772/65195

Abstract

The extent of herbivory in lizards is influenced by several factors. Plant tissues are more difficult to digest than invertebrates due to the presence of cellulose. Thus, so many lizards exhibit carnivorous diet. Nevertheless, some species consume vegetables. Essentially herbivorous diet occurs in about 3% of lizards, while most omnivores add plants in their diets. Omnivorous species tend to eat more fruits, flower, and nectar, because they are easier to digest and provide more nutrients than leaves, which are rich in cellulose. The main factors influencing the consumption of plant material are related to the habitat of the species. Insular and arid environments favor the consumption of plants because such locations have low amount of arthropods available and present water scarcity. It is also possible to observe ontogenetic changes in the lizard's diet, in such a way that young individuals consume only invertebrates, whereas the adults supplement their diet with plant material. When consuming fruits and nectar, lizards become potential dispersers and pollinators. In this sense, some studies have already corroborated seed dispersal and pollination events by lizards. In islands where other species are absent, these interactions are essential for the maintenance of communities.

Keywords: herbivory, lizard ecology, ontogenetic, seed dispersion, diet, omnivory

1. Introduction

Due to lower consumption of vegetables by lizards, often the groups are overlooked in studies that discuss herbivory. However, it is important to discuss this matter in order to obtain a better understanding with regard to lizards/plant interaction, inasmuch as there are determinants physiological characteristics for the occurrence or absence of such interaction. Considering that the plants compose the diet of lizards, it is worth emphasizing the ecological importance of this relationship, which composes the trophic chain.

A few species of living lizards currently present are essentially herbivores, especially when compared with the extensive fauna of herbivorous reptiles of the Mesozoic [1]. Regarding diet, most lizards consume small animals and rarely plant material. On the other hand, there are omnivores species and some exclusively herbivores [2]. Accordingly, many factors can influence the consumption of plants, such as seasonal availability of food items, digestibility, and components from the consumed plants [3].

The nutrient assimilation capacity in herbivores is less than in omnivores, while such capacity is lesser in omnivores than in carnivores [1]. The low nutrient assimilation capacity in lizards reduces growth rates and reproductive capability, causing a reduction in egg production by females [2, 4]. Thus, the consumption of plants can make them more susceptible to predation, influencing in the adaptive radiation of the species.

2. Vegetables in composition of the lizard's diet

An herbivorous diet requires adaptations, physiological or behavioral, for the digestion of cellulose. Such adaptations may be: specialized dentition, elongated intestines, colic valves, intestinal flora, and thermoregulation to maintain high body temperature [5], inasmuch as gut fermentation, which is necessary for plant digestion, requires prolonged periods of high body temperature [6]. Every part that makes up a plant requires different adaptations for digestion because the parts vary significantly in their structure and composition. Leaves and stems are more difficult to digest than nectar, pollen, flowers, and fruits, due to the greater amount of cellulose present.

The consumption of plant material by lizards is less frequent than in other groups, such as mammals and birds. Many species consume vegetables, but they are considered omnivorous (capable of metabolizing different food items) and the majority of their diet is composed of small animals, consisting essentially of arthropods. Less than 3% of lizard species are considered essentially herbivorous [1, 2, 4, 7]. Lizards are considered strictly herbivorous only if more than 90% of its diet is composed of plant materials [3]. The ectothermy[1] is one of the reasons for the lower consumption of vegetables by the lizards, compared to mammals and birds, insofar as the temperature variations according to the environment difficult digestion of these materials. For the occurrence of the digestion of plant material, a long time of thermoregulation is necessary, because at low temperatures the absorption of nutrients is compromised, reducing the energetic efficiency. On the other hand, a long exposure time for thermoregulation increases the risk of predation. As an example, it can be cited that the lizard *Dipsosaurus dorsalis* (Baird and Girard, 1852), a herbivorous lizard from the desert, maintains higher temperatures than other iguanids insectivorous lizards and is active for a long period of the day [8].

Regarding this issue, for some species, to increase the body temperature is not the main mechanism used for digestion of plant material. The lizard *Cnemidophorus murinus* (Laurenti,

[1] An ectothermic animal vary its body temperature according to the ambient temperature but control this variation by behavioral methods. For example, the lizards remain exposed to the sun in the morning, and they hide from the sun during the rest of the day in order to prevent overheating.

1768) keeps the body temperature similar to other teiids lizards but it remains active much longer to facilitate the digestion of plant [9], in that way the body temperature is maintained constant at about 37°C, predominantly during the day [6].

Since plant tissues are more difficult to digest than animal tissue, lizards require mechanisms that facilitate the digestion. Moreover, birds and mammals that are exclusively herbivores have high energy efficiency due to temperature maintenance and digestive symbiotic associations [4].

Then, it raises the following question: what are the factors that allow some species of lizards feed on plants and not others? The first studies that have sought to elucidate this query, consider the size of the animal as decisive in choosing the type of diet. Theoretically, harder items need greater strength and efficiency in chewing, and to have a larger and more robust head, allowing a stronger bite [10]. Campos, in a work about the lizards of *Microlophus* genus, found differences in cranial morphology associated with their diet: species that consume plant material have larger and wider skulls than the insectivorous ones. In this sense, large lizards could employ greater strength in their jaws, producing a better grinding of the food. For a time, it was assumed that only large lizards would be able to reduce plant material into digestible portions [7]. Several recent studies have shown that smaller species are able to consume plant material. Nevertheless, the adaptations which allow these species consume plants are lesser known [5].

Besides the initial process of tissues breaking by chewing, for digestion of more resistant materials, it is necessary a more elongated digestive tract and, the presence of symbiotic relationships. More elongated intestines, as well as chewing ability, are related to the body size of the animal. It is known that most herbivorous lizards belong to Iguanidae family. This is one of the few groups of lizards known which have expertise to digest plant material. The iguana intestine contains colic valves in order to increase the area of absorption and residence time of food in the intestines as well as intestinal flora suitable for cellulose degradation [3].

In a manner contrary to the trend related to body size, *Liolaemus lutzae* (Mertens, 1938), a species of small size, consumes large amounts of plant material, presenting evidence of capacity to cut leaves to consume them [11]. The same author concluded that the consumption of leaves by the species does not occur indiscriminately, inasmuch as the most common plant species in the stomach contents are less frequent in the studied area, suggesting that there is selection of the items to be consumed by their qualitative properties: amount of cellulose, sugars, fibers, tannins, and other components of the plant.

Omnivores Lizards, which include plant materials in their diets, tend to consume the softer parts of the plant, such as flowers and fruits containing large amounts of lipids, carbohydrates, sugar, and protein [12].

The selection of certain plant parts is also observed in other species, such as *Tropidurus torquatus* (Wied-Neuwied, 1820), species of omnivorous diet which also consumes plant material, tends to consume fruits, since they have a good digestibility [13]. Similarly, *Tupinambis teguixin* (Linnaeus, 1758) in a population of eastern Chaco, Argentina, presented plant material as its main diet composition (over 60%), in which most were fruits [14], even though it is a species of large size. Therefore, there is not an only factor, but a set of factors that influence the herbivory in the lizards.

Leiocephalus carinatus (Gray, 1827), another omnivorous species, has on your diet 47% of plant material. Kircher *et al.* [15] reported the use of *Ipomoea pes-caprae* flowers by adults of the species, in the Cayman Islands, Cuba.

2.1. Food preference in herbivorous and omnivorous lizards

Omnivorous species, as well as the herbivorous ones, can select parts of plants rich in nutrients and low cellulose content to consume, according to their physiological demand or digestive adaptations. Lizards that consume plant material in small quantities, or that do it occasionally, choose parts easier to digest or that which do not require specific adaptations to digestion. Lizards of omnivorous diet consume fruits and flowers intentionally when their preys are scarce. Conversely, the consumption of fragments of plant material by carnivorous animals occurs accidentally [3].

Some lizard species have food preferences according to the presence or absence of certain components. *Dicrodon guttulatum* (Duméril and Bibron, 1839), from Teiidae family, prefers the *Prosopis pallida* tree. Velásquez *et al.* [16], examining the *D. guttulatum* diet, in the *P. pallida* absence, found predominantly leaves of *Acacia* sp. and fruits from *Scutia spicata* and *Capparis* sp. In a study carried out by Leeuwen et al. [5] about the diet of the same lizard species in an environment with *P. pallida*, it was detected mainly this plant species in stomach contents and rarely other plants, even if they were available in the environment.

The presence of water in the fruit is also an important factor for consumption. Figueira et al. [17], investigating the interaction between *T. torquatus* with *Melocactus violaceus* (Cactaceae) reported the consumption of fruits always that they were available. These fruits have elevated water content and low sugar levels, indicating that the plant can be important to supply the need for water.

Although less frequent, some studies have looked for to understand how the lizards detect or choose vegetables to be consumed. Vasconcellos-Neto et al. [18], using models of artificial fruit, concluded that *T. torquatus* consume fruits that have higher color contrast in relation to the substrate or plant where they are, as well as they prefer conic fruits to other forms. It should be noted that the species has accurate visual perception, used for catching prey. As previously mentioned, *T. torquatus* consumes fruits of *M. violaceus*, which is a cactus whose fruits are conical, colorful, and close to the ground. That fact supports the hypothesis that lizards can to use visual perception to the fruit location and for identification of the ideal shape for the extrusion process [17].

The chemical perception ability for plant identification is correlated with the evolution of herbivory in Scleroglossa [3]. For Iguania lineage there are records of using of chemical samples collected by language in *D. dorsalis* [19]. In iguanas, the chemical discrimination of food has a strong correlation with the evolution of herbivory [20]. The use of chemical perception is important in the consumption of plant material, as these food items are motionless, making difficult the visual identification. Similarly, if different parts of the plant and different species have distinct compositions, the act of identifying the food before eating increases energy efficiency and also may avoid the consumption of species with toxins [3].

Although small fruits tend to be more easily handled and eaten by frugivorous, studies with lizards have not confirmed this trend. Rodríguez-Pérez and Traveset [21] studied the interaction between *Daphne rodriguezii*, an endemic shrub from Menorca Island (in Spain), with the lizard *Podarcis lilfordi* (Günther, 1874), which is its disperser. In such study the seeds size was not an important factor for the selection of fruits, because seeds are not predictors of the amount of pulp mass.

The chameleon *Furcifer oustaleti* (Mocquard, 1894) consume plant parts using different mechanism of collection if compared with animal prey. The capture of arthropods occurs through the projection of his tongue, while the consumption of fruit, for example, takes place by direct collection of the jaws. The mechanism indicates that the individuals identify the food item before consuming them and, since there is no need to capture vegetables items, the chameleons opt for direct collection, saving energy spent on the tongue projection [22].

Benítez-Malvido et al. [23] in an investigation about seed dispersal by *Iguana iguana* (Linnaeus, 1758), have observed that individuals of the species select the items from their diet, consuming fleshy fruits. The reason is that fleshy fruits have higher amount of water, sugars and less material of hard digestion. In the same study it was observed that puppies of *I. iguana* have difficulty eating large fruits or coriaceous ones because of the reduced body size.

It is worth mentioning that the presence of flowers and fruits in the diet of lizards is strongly influenced by the period in which the study was developed, due to seasonal availability of these items [24]. For that reason, the relative importance of consumption of plant material is dependent on the period of the sample, which limits discussions about the subject.

Some lizards choose to consume nectar or pollen in order to have a higher concentration of nutrients. That enables individuals pollinate plants, since they visit different flowers. The consumption of nectar for lizards is rarer, but there are some records. Although not common, such interaction is important, as it allows for pollination events. And, since the consumption occurs mainly on islands [25], environments that often lack pollinators, the lizards are fundamental in the maintenance of communities.

Geckos of the *Hoplodactylus* genus, in New Zealand, consume pollen from native plant species [26]. Geckos increase the consumption of flowers as enhance the viscous nectar (53% sugar). *P. lilfordi* are also potential pollinators of flowers belonging to the *Euphorbia dendroides* species, at island environments [27]. Furthermore, there are records of pollen consumption for the lacertid lizards *Gallotia simonyi* (Steindachner, 1889), *P. lilfordi*, and *Podarcis pityusensis* (Bosca, 1883) and the gecko *Rhacodactylus auriculatus* (Bavay, 1869) [3, 27].

2.2. Environmental factors

Factors related to the habitat of species may have direct influence on the consumption of vegetables by lizards, such as the availability of prey, aridity [3] and island environments [15].

The evolution of plant consumption may be favored by habitat factors that reduce the availability of prey [3]. It is observed on islands with low predation rates [4]. Another factor that likely influences in vegetable consumption is the intraspecific competition. When the density of the population increases, reducing the availability of prey, the feeding of alternative items favors the maintenance of the species.

Herbivorous in lizards evolves more commonly in species that occur in small islands, and appears to result from a lower abundance of arthropod preys available in these habitats [15]. An evidence is that when the diets of two populations of *T. torquatus* were compared, one continental and another insular, they were the same. This is likely because the availability of arthropods in both environments was similar [28].

In addition, as in islands, the species diversity is smaller, the occurrence of predators is reduced; therefore, the risk of exposure to the sun for thermoregulation is also reduced. Considering that, in that case, the lizards are able to maintain higher body temperatures, the energy efficiency rises, becoming advantageous the consumption of vegetables.

Fossil records of *Gallotia* clade, a group of lacertid lizards that inhabit the Canary Islands, shows that island environments are crucial in the development of herbivory, since even the individuals which are primitively of large body, the herbivory just has developed after the colonization of the islands [29]. Before the colonization they were carnivores. This study highlights the effect of abiotic factors on the ecology of the species.

However, all the evidence that island environments favor the development of herbivory are hypothesis, they have not been tested [3].

There are evidences that an arid environment favors the development of herbivory because of seasonal and unpredictable scarcity of prey. On the other hand, it is not possible to assess the effect of this factor on diet in an isolated way, since many islands are arid, becoming impossible to isolate the factors.

Anyway, the main factor directly related to the consumption of plants, predominantly in omnivorous species, is the availability of prey. About 80% of species respond to that factor.

The availability of prey is so decisive for the establishment of populations of insectivorous species that lizards of Liolaemidae family have a tendency to herbivory, related with the climate, which is contrary to the other groups. Espinoza et al. [30], in an article concerning the evolution of herbivory in individuals belonging to that family, concluded that vegetable consumption is directly correlated with cold climates while most herbivores live in habitats with hot and dry climate. Herbivorous tropical lizards (*Iguana, Ctenosaura,* and *Amblyrhynchus*) live in environments where the temperature remains high in most of the day.

2.3. Distribution of vegetable consumption by taxon

Living lizards are divided into two large lineages: the Iguania (e.g., iguanas and chameleons) and Scleroglossa (all other lizards). These groups have pronounced differences in foraging behavior: Iguania presents ambush foraging, consuming mainly insects, with some herbivores and omnivores individuals, which is a behavior derived from ambush forage. On the other hand, Scleroglossa includes mostly individuals with active foraging, a few ambush foragers, and a small amount of herbivores and omnivores [20].

The consumption of plant material occurs in the families of both lineages: Iguania and Scleroglossa. In Iguania it is present in almost every family with known data. Except in Chamaeleontidae and Crotaphytidae, the vegetable diet is of universal occurrence in Iguanidae (occurs in all species) and frequent in the species of the Tropiduridae family. In Scleroglossa the herbivory is absent in Pygopodidae, Eublepharidae, Gymnophthalmidae, Cordylidae, and most families of Anguimorpha [3].

Despite the lack of known robust phylogenetic relationship in the lizards, which makes difficult to trace the exact epoch that occurred the vegetable consumption evolution in these reptiles, the discovered patterns allow to infer that such evolution happened in several periods of their evolutionary tree.

In the following, some comments about the main families are presented, for which studies have been carried out concerning herbivory.

2.3.1. *Iguanidae*

Most herbivorous lizards belong to the Iguanidae family. The majority of species are strictly herbivorous. Others present ontogenetic changes, consuming plant material when they are adults. This family also has folivorous feeders. It is worth remembering that some species have their adaptations proper to the consumption of plant material, as previously mentioned. Another fact is that the herbivory is present in the common ancestor of Iguanidae [3].

2.3.2. *Tropiduridae*

In Tropiduridae family, the consumption of plant material is well known. Most genera are omnivores. And the degree of consumption varies considerably between genera and species. Many studies have shown data about the Tropidurus genus, stating that its diet has ontogenetic, seasonal and geographical variations [13, 31], as well as it reflects to local availability of foods [28]. *Liolaemus lutzae* also presents ontogenetic change in its diet [11].

2.3.3. *Agamidae*

The Agamidae family, which comprises lizards known as Australian Water Dragon, has some omnivores genera and only two herbivores genera while *Hydrosaurus pustulosus* is an exclusively folivorous species [3]. The other species that include plant material in their diet tend to consume the parts easier to digest, such as fruit and leaves.

2.3.4. *Gekkonidae*

Many species of the Gekkonidae family consume small amounts of plant material. Some geckos from New Zealand have omnivorous diet with consumption of different parts of the plant. Wotton [24] described the diet of *Mokopirirakau granulatus* (Gray, 1985), while for *Naultinus grayii* (Bell, 1843) it was reported by Whitaker [26].

2.3.5. Teiidae

Most species of Teiidae family are carnivorous. The species considered omnivorous belong to *Cnemidophorus, Tupinambis* [3] and *Ameiva* genera [32]. *Tupinambis teguixin* includes in its diet all parts of the plant, with predominance of fruits [14].

2.4. Ontogenetic and seasonal variations in vegetable consumption

The diet of lizards presents variations in response to different factors. Some species show changes between the sexes, possibly because of morphological differences, as in many species the males have larger body size than females. Other studies have indicated the seasonal variation in the diet. The species with such variation, in general, live in environments where there is rainfall variation over the year. However, a large part of the diet variation in lizards is explained by the availability of food in their living area. Considering that the main plant components in the diet of lizards are fruits and flowers, it is expected a seasonal influence on the diet [24].

There is still, in a long term, ontogenetic changes in the diet of lizards, when young individuals have a different diet in comparison to the adults from the same species. This fact occurs due to the difference in energetic demands during the growth of individuals. Most lizards which are omnivorous and also consumes vegetables, as young they feed basically on insects. Only *I. iguana* has an exclusively herbivorous diet, both in adults and young ones [25].

The evidences of ontogenetic change in the consumption of vegetables can be observed if it is considered the quantity of vegetables consumed or in relation to the plant parts that make up the diet such as leaves, fruits, flowers, or nectar. Since each component has a different proportion of protein, water, fiber, glycides, and cellulose (the most difficult component to digest). During the period, the demand for protein is higher thus, a diet plant-based would not supply the energy needs. Therefore, puppies tend to look for food items with higher amounts of protein [3].

As adults, certain species begin to add larger amount of plant material in the diet. There are several reasons for that. It may be the greater availability of these items, since in certain environments, such as islands, the availability of arthropod is reduced; presence of water in the fruits (in arid environments is an essential feature for maintaining the species metabolism); more facility of handling when compared to prey, inasmuch as in order to capturing small invertebrates it is necessary a higher energetic spent than the consumption employed only in collecting fruits. For example: reduced competition, as youth and adults consume different food items, they will not compete for the same niche, and nutritional content.

The ontogenetic change in alimentation is well marked in omnivorous species despite being also observed in essentially herbivorous species. *Microlophus thoracicus* (Tschudi, 1845), an omnivorous species, presents ontogenetic change in its diet, consuming insects as juvenile, adding vegetables when adult [10].

Rocha [11] carried out a study about ontogenetic changes in *L. lutzae* diet in Barra do Baricá, coast of Rio de Janeiro (Brazil). Despite its small size during adulthood (60–80 mm), nearly half of the diet of the species was composed by vegetables material, and the presence of these items increased according to increase of age and body size of the individuals. This is reinforced by the fact that lizards with snout-vent-length (SVL) smaller than 38 mm consumed only arthropods. Such trend was present in both sexes. However, males tend to consume more plant material than females. The low availability of arthropods in beach environments and abundant presence of shrubs make the ontogenetic change an advantage in terms of digestive efficiency. Then, the species begin to explore new niche, limiting the consumption of arthropods by adults in order to not competing intraspecifically with juveniles.

Populations of *T. torquatus*, which inhabiting the same environment previously mentioned, have showed behavior similar to the *L. lutzae*, increasing gradually their diet with plant material during growth, and when adult they have almost half of the diet consisting of plant material [13]. The two papers, Rocha and Fialho et al., highlight the importance of environmental factors in the composition of the lizard's diet, leading the species to consume items that theoretically would be disadvantageous due to the difficulty of digestion, but that are more accessible.

In environments with seasonality marked by precipitation difference, as well as in tropical environments with dry and rainy seasons (Barra de Maricá), the volume of rain is decisive for the availability of arthropods and also for the plants cycle. Since the more rain the more availability of arthropods. Thus, the diet will vary according to the availability of food in each station. It should be noted also that both species (*L. lutzae* and *T. torquatus*) are omnivorous, opportunistic predators and their diets reflect the availability of food items in the habitat.

3. Lizards as a seed disperser

In order to consider an animal that consumes fruits as a disperser, it is necessary to validate quantitative and qualitative factors. Quantitative factors depend on the number of seeds consumed, while the qualitative ones depend on the location in which the seed is deposited and the effect of the passage through the digestive tract on seed germination [33–35]. The treatment that the seeds receive when they are consumed directly influences the capacity and the speed of germination [21]. In the same way that the distribution pattern of seeds on the microhabitats is a crucial aspect of the dispersion quality [36]. Inasmuch as different microhabitats provide different conditions (illumination, humidity, substrate characteristics, etc.), causing alterations in the rates of germination and seedling survival.

Similarly to the studies of herbivores, research about dispersion syndrome also focuses mainly on birds and mammals. Studies concerning that syndrome in reptiles are less common [23]. However, this group plays an important role, especially species which inhabit arid environments and islands.

Information on seed dispersal by lizards are known for: geckos in New Zealand islands [24, 26]; lizards in various environments in Brazil [31, 37], in the Mexico (Benítez-Malvido et al.), in island of the Western Mediterranean [21, 38] and iguanas in the Galapagos Islands [39] and dry forests of Costa Rica [40].

Many studies have shown a positive effect of the passage through the digestive tract of lizards on the ingested seeds. However, this effect varies with the species consumed.

The environment analyzed by Wotton [24], an island in New Zealand where the gecko *Woodworthia maculatus* lives (Gray, 1845), has reduced populations of birds and mammals that may disperse the seeds, and compete by the consumption of the fruits. In this sense, the gecko has fundamental importance for the maintenance of the local communities, due to its dispersal potential.

The seed dispersal by lizards is peculiar because, even though being characterized by local events or short distances, it has fundamental importance due to the tendency to eat fallen fruit, inaccessible to other vertebrates like birds [24].

In a paper about seed dispersal by *T. torquatus*, Pietczak *et al.* [37] have found that the seeds deposition of *Chomelia obtusa* (species consumed by the lizard at the studied site) usually occurs in a short distance from the mother plant, around five meters. But, despite the fact that the average dispersal distance is not too long, the species is benefited. The reason for this is that, according to Chapman and Chapman [41], seeds dispersed even in short distances germinate better than ones under the parent plant. The dispersion of seed in locations similar to that of natural occurrence of the species indicates favorable conditions for the germination. In this sense, Pietczak et al. verified that the population of *T. torquatus* studied merely remained in the areas around a rocky outcrop, depositing the seeds on the edges and clefts of the rocks. Therefore, the places where the lizard deposited the seeds favored the seed germination and seedling development.

Rodríguez-Pérez and Traveset [21], in a paper about the interaction of the bush *D. rodriguezii* and its disperser, namely, the lizard *P. lilfordi*, stated that the passage through the digestive tract neither increased nor decreased the germination capacity. Nevertheless, the passage through the digestive tract appeared to have caused a reduction in size of the seed, with an action on the coating thickness. Traveset [33] indicates that the reduction of the coating serves as a scarifying process, increasing the permeability of the seed and thus favoring the germination. However, such permeability was not evaluated in detail by Rodríguez-Pérez and Traveset.

Once again, regarding the work of Rodríguez-Pérez and Traveset [21], it was observed that seedlings of seeds deposited under the mother plant had lower survival rates compared to the seeds dispersed. Accordingly, even though the passage through the digestive tract of lizards does not significantly increase the germination rate, the deposition pattern of seeds increases the viability of seedlings survival. Thus, the action of lizards is characterized as a disperser. Another paper, studying the same disperser, has obtained similar conclusions, in which the effect of ingestion had neutral results on germination rates, but the seed deposition favored the development of seedling [38]. In this case in particular, since *P. lilfordi* é is the

http://dx.doi.org/10.5772/65195

only disperser of *D. rodriguezii*, such interaction is essential for the maintenance of the species.

4. Concluding remarks

There are some studies that explain the consumption of vegetables by lizards. For a while it was considered the body size as a decisive factor for herbivory. It was thought that only large lizards were able to consume plant material. More recent publications, with small- and medium-sized species, have shown that there are various reasons for some species consume vegetables and others do not.

In relation to the anatomy, physiology and behavior of the species, it is necessary certain adaptations to consume vegetables, such as: specialized dentition, elongated intestines, colic valves, intestinal flora, and thermoregulation to maintain high body temperature. The Iguanas are the lizards with the most elaborate adaptations to herbivory.

Many lizards which have omnivorous diet consume plant material. For such a species of lizards, changes in their diet are strongly related to environmental factors. Vegetable consumption by lizards is more likely to occur in insular and arid environments, as well as in areas with reduced predators. But the deciding factor associated with the addition of vegetable items is the availability of prey. Insular and arid environments have lower availability of arthropods, which is the main food item of most lizards. Nevertheless, more studies are needed to isolate each of the following factors: insularity, aridity and availability of prey.

Environments with prey scarcity also favor the ontogenetic changes in the diet. If there are not enough arthropods for population maintenance, it is more beneficial that youth individuals maintain a diet richer in protein (arthropods) and adults change their diets for items with higher availability (vegetables). Since many species of omnivorous diet do not present physiological or anatomical adaptations to the digestion of plant tissue, they choose the parts of the plants easier for the digestibility, like fruits, flowers, and nectar.

During the consuming of fruits and nectar, some lizard species can disperse seeds and pollinate flowers. The main contribution of the lizards in the dispersal of seeds is on the deposition pattern of seeds. More studies are needed in order to investigate the dispersion syndrome in other species and draw a general profile of seed dispersal by lizards.

The pollination by lizards is rare, but very relevant, because the occurrence of such events is carried out on islands. These environments may have reduced diversity of other animal groups and are isolated. The interactions of the plants with lizards are fundamental to the maintenance of species on islands.

Therefore, the research conducted so far about herbivory by lizards have shown interesting results, but many hypotheses have yet to be formulated and tested.

Author details

Carolina Pietczak[1*] and Lucas R. Vieira[2]

*Address all correspondence to: carolpietczak@gmail.com

1 Instituto Federal de Educação, Ciência e Tecnologia Farroupilha, Santo Augusto, Brazil

2 Instituto Federal de Educação, Ciência e Tecnologia Catarinense, Concórdia, Brazil

References

[1] Pough H. Lizard energetics and diet. Ecology. 1973; 54: 837–844. DOI: 10.2307/1935678

[2] Iverson JB. Lizards as seed dispersers. Journal of Herpetology. 1985; 19: 292–293.

[3] Cooper WE, Vitt LJ. Distribution, extent, and evolution of plant consumption by lizards. Journal of Zoology. 2002; 57: 487–51. DOI: 10.1017/S0952836902001085

[4] Szarsky H. Some remarks on herbivorous lizards. Evolution. 1962; 16: 529.

[5] Leeuwen J, Catenazzi A, Holmgren M. Spatial, ontogenetic, and sexual effects on the diet of a teiid lizard in arid South America. Journal of Herpetology. 2011; 45(4): 472–477.

[6] Vitt LJ, Caldwell JP, Sartorius SS, Cooper WE, Baird TA, Baird TD, Pérez-Mellado V. Pushing the edge: Extended activity as an alternative to risky body temperatures in an herbivorous teiid lizard (*Cnemidophorus murinus*: Squamata). Functional Ecology. 2005; 19: 152–158.

[7] Sokol OM. Herbivory in lizards. Evolution. 1967; 21: 192–194.

[8] Pianka ER. Comparative ecology of two lizards. Copeia. 1971; 1: 129–138.

[9] Vitt JL. Shifting paradigms: Herbivory and body size in lizards. PNAS. 2004; 101(48): 16713–16714.

[10] Campos KST. Evidence of adaptive evolution in the cranial morphology of Tropidurid lizards from coastal Peru. Herpetology Notes. 2016; 9: 47–53.

[11] Rocha CFD. Ontogenetic shift in the rate of plant consumition in a tropical lizard (*Liolaemus lutzae*). Journal of Herpetology. 1998; 32(2): 274–279.

[12] Siqueira CC, Kieffer MC, Van SM, Rocha CFD. Plant consumption in coastal populations of the lizard *Tropidurus torquatus* (Reptilia: Squamata: Tropiduridae): How do herbivory rates vary along their geographic range? Journal of Natural History. 2011; 45(3–4): 171–182. DOI: 10.1080/00222933.2010.520826

[13] Fialho RF, Rocha CFD, Vrcibradic D. Feeding ecology of *Tropidurus torquatus*: Ontogenetic shift in plant consumption and seasonal trends in diet. Journal of Herpetology. 2000; 34(2): 325–330.

[14] Mercolli C, Yanosky A. The diet of adult *Tupinambis teguixin* (Sauria: Teiidae) in the eastern chaco of Argentina. Herpetological Journal. 1994; 4: 15–19.

[15] Kircher BK, Robinson CD, Johnson MA. Herbivory in the northern curly-tailed lizard (*Leiocephalus carinatus*). Caribbean Herpetology. 2014; 50: 1–2.

[16] Velásquez LP, Estraver WZ, Pinedo AT, Chinchay LP. Hábitos alimentarios de Dicrodon guttulatum, "caña" (Squamata:Teiidae) en Garrapón, Paiján. Arnaldoa. 2007; 14: 283–291.

[17] Figueira JEC. Saurocory in *Melocactus violaceus* (Cactaceae). Biotropica. 1994; 26(3): 295–301.

[18] Vasconcellos-Neto J, Souza ALT, Guimarães MM, Farias DM. Effects of color, shape and location on detection of cactus fruit by a lizard. Journal of Herpetology. 2000; 34(2): 306–309.

[19] Cooper WE Jr, Alberts AC. Responses to chemical food stimuli by an herbivorous actively foraging lizard, *Dipsosaurus dorsalis*. Herpetologica. 1990; (46): 259–266.

[20] Cooper WE. Correlated evolution of herbivory and food chemical discrimination in iguanian and ambush foraging lizards. Behavioral Ecology. 2003; 14(3): 409–416.

[21] Rodríguez-Pérez J, Traveset A. Seed dispersal effectiveness in a plant–lizard interaction and its consequences for plant regeneration after disperser loss. Plant Ecology. 2010; 207: 269–280.

[22] Takahashi H. Fruit feeding behaviour of a chameleon *Furcifer oustaleti*: Comparison with insect foraging tactics. Journal of Herpetology. 2008; 42: 760–763.

[23] Benítez-Malvido J, Martínez-Ramos. Impact of florest fragmentation on undeestory plant species richness in Amazonia, Conservation Biology. 2003; 17: 389–400.

[24] Wotton DM. Effectiveness of the common gecko (*Hoplodactylus maculatus*) as a seed disperser on Mana Island, New Zealand. New Zealand Journal of Botany. 2002; 40: 639–647. DOI: 0028-825X/02/4004-0639

[25] Olesen JM, Valido A. Lizards as pollinators and seed dispersers: An island phenomenon.Trends in Ecology and Evolution. 2003; 18(4): 177–181.

[26] Whitaker AH. The roles of lizards in New Zealand plant reproductive strategies. New Zealand Journal of Botany. 1987; 25(2): 315–328. DOI: 10.1080/0028825X.1987.10410078

[27] Pélez-Mellado V, Traveset A. Relationships between plants and mediterranean lizards. Natura Croatica. 1999; 8: 275–285. DOI: 598.112. 581.162.3/591.13

[28] Dutra GF, Siqueira CC, Vrcibradic D, Kiefer MC, Rocha CRD. Plant consumption of insular and mainland populations of a tropical lizard. Herpetologica. 2011; 67(1): 32–45. DOI: http://dx.doi.org/10.1655/HERPETOLOGICA-D-09-00009.1

[29] Cernanski A, Klembara J, Smith KT. Fossil lizard from central Europe resolves the origin of large body size and herbivory in giant Canary Island lacertids. Zoological Journal of the Linnean Society. 2016; 176: 861–877. DOI: 10.1111/zoj.12340.

[30] Espinoza RE, Wiens JJ, Tracy R. Recurrent evolution of herbivory in small, cold-climate lizards: Breaking the ecophysiological rules of reptilian herbivory. PNAS. 2004; 101(48): 16819–16824.

[31] Fialho RF. Seed dispersal by a lizard and treefrog – Effect of dispersal site on seed survivorship. Biotropica. 1990; 22(4): 423–424.

[32] Vitt LJ, Colli GC. Geographical ecology of a neotropical lizard: *Ameiva ameiva* (Teiidae) in Brazil. Canadian Journal of Zoology. 1994; 72: 1986–2008.

[33] Traveset A. Effect of seed passage through vertebrete frugivores' guts on germination: A review. In: Perspectives in Plant Ecology, Evolution and Systematics. 1998; 1/2: 151–190.

[34] Jordano P, Galetti M, Pizo MA, Silva W. Ligando frugivoria e dispersão de sementes à biologia da conservação. In: Rocha CFD, Bergallo HG, Sluys MV, Alves MAS. Biologia da conservação: Essencias. São Paulo: RIMA. 2006; p. 411–435.

[35] Schupp EW, Jordano P, Gómez JM. Seed dispersal effectiveness revisited: A conceptual review. New Phytologist. 2010; 188: 333–353.

[36] Celedón-Neghme C, Traveset A. Contrasting patterns of seed dispersal between alien mammals and native lizards in a declining plant species. Plant Ecology. 2013; 214: 657–667. DOI: 10.1007/s11258-013-0197-7

[37] Pietczak C, Arruda JLS, Cechin SZ. Frugivory and seed dispersal by *Tropidurus torquatus* (Squamata: Tropiduridae) in southern Brazil. The Herpetological Journal. 2013; 23(2): 75–79.

[38] Castilla MA. The lizard *Podarcis lilfordi* as a potential disperser of the Solanaceae plant *Lycopersicon esculentum*: Can legitimate dispersers indirectly promote plant invasions? Munibe (Ciencias Naturales-Natur Zientziak). 2009; 57: 185–194.

[39] Hendrix LB, Smith SD. Post-eruption revegetation of Isla Fernandina, Galapagos (Ecuador). II. National Geographic Research. 1986; 2: 6–16.

[40] Traveset A. Post-dispersal predation of *Acacia farnesiana* seeds by *Stator vachelliae* (Bruchidae) in Central America. Oecologia. 1990; 84(4): 506–512. DOI: 10.1007/BF00328167

[41] Chapman CA, Chapman LJ. Survival without disperser: Seedling recruitment under parent. Conservation Biology. 1985; 9: 675–678.